Springer Theses

Recognizing Outstanding Ph.D. Research

Aims and Scope

The series "Springer Theses" brings together a selection of the very best Ph.D. theses from around the world and across the physical sciences. Nominated and endorsed by two recognized specialists, each published volume has been selected for its scientific excellence and the high impact of its contents for the pertinent field of research. For greater accessibility to non-specialists, the published versions include an extended introduction, as well as a foreword by the student's supervisor explaining the special relevance of the work for the field. As a whole, the series will provide a valuable resource both for newcomers to the research fields described, and for other scientists seeking detailed background information on special questions. Finally, it provides an accredited documentation of the valuable contributions made by today's younger generation of scientists.

Theses are accepted into the series by invited nomination only and must fulfill all of the following criteria

- They must be written in good English.
- The topic should fall within the confines of Chemistry, Physics, Earth Sciences, Engineering and related interdisciplinary fields such as Materials, Nanoscience, Chemical Engineering, Complex Systems and Biophysics.
- The work reported in the thesis must represent a significant scientific advance.
- If the thesis includes previously published material, permission to reproduce this must be gained from the respective copyright holder.
- They must have been examined and passed during the 12 months prior to nomination.
- Each thesis should include a foreword by the supervisor outlining the significance of its content.
- The theses should have a clearly defined structure including an introduction accessible to scientists not expert in that particular field.

More information about this series at http://www.springer.com/series/8790

Swee Leong Sing

Selective Laser Melting of Novel Titanium-Tantalum Alloy as Orthopaedic Biomaterial

Doctoral Thesis accepted by
the Nanyang Technological University,
Singapore

Springer

Author
Dr. Swee Leong Sing
Singapore Centre for 3D Printing,
School of Mechanical and
Aerospace Engineering
Nanyang Technological University
Singapore

Supervisors
Dr. Wai Yee Yeong
Singapore Centre for 3D Printing,
School of Mechanical and
Aerospace Engineering
Nanyang Technological University
Singapore

Dr. Florencia Edith Wiria
Singapore Institute of Manufacturing
Technology
Agency for Science, Technology
and Research (A*STAR)
Singapore

ISSN 2190-5053 ISSN 2190-5061 (electronic)
Springer Theses
ISBN 978-981-13-2723-0 ISBN 978-981-13-2724-7 (eBook)
https://doi.org/10.1007/978-981-13-2724-7

Library of Congress Control Number: 2018955911

This Springer imprint is published by the registered company Springer Nature Singapore Pte Ltd.
The registered company address is: 152 Beach Road, #21-01/04 Gateway East, Singapore 189721, Singapore

Supervisors' Foreword

We are very pleased that Springer is publishing Swee Leong's dissertation titled *Selective Laser Melting of Novel Titanium-Tantalum Alloy as Orthopaedic Biomaterial*. It is an honour that this research work is recognised as impactful and outstanding in the global arena. Selective laser melting (SLM) is a powder bed fusion additive manufacturing (AM) technique that is capable of fabricating complex functional three-dimensional (3D) metal parts of high relative density with the complete melting and fusion of powders. There are many research works reported on SLM. The key unique novelty of this work is that Swee Leong successfully identified a new research dimension in AM and utilised SLM as an ultra-precise 3D tool in material formulation and formation. SLM has the potential in expanding the materials library by in situ formation of alloys that were previously difficult to achieve by using metal powder mixtures. This new capability adds a new dimension to the advantages of enabling customisation according to the application requirements. This dissertation used titanium-tantalum as a proof of concept for this in situ alloying method.

Titanium-tantalum (TiTa) is a potential material for biomedical applications due to its high strength-to-modulus ratio. However, it is still not widely used due to the difficulties in obtaining this alloy. Due to its versatility in processing metallic materials and good results obtained from commercially pure titanium (cpTi) and Ti6Al4V, Swee Leong has chosen SLM as the process to tackle this problem. His research developed TiTa as a potential material for biomedical applications and paved the way for better understanding and control of the SLM process in porous lattice structure fabrication through statistical modelling.

From the 3D printing process perspective, Swee Leong studied TiTa formation to understand the forming mechanism and the effect of SLM processing parameters on the resulting density and macrostructure of the parts. The processing window and optimised parameters based on optimum relative density achieved are then presented.

In addition, Swee Leong examined the resulting microstructure of TiTa parts to gain extensive knowledge behind the material formation by SLM. The SLM TiTa parts exhibited a microstructure characterised by homogenous β titanium and tantalum matrix with randomly distributed tantalum particles. It is found that SLM TiTa parts have lower Young's modulus and comparable strength to Ti6Al4V and cpTi.

It is tedious and time-consuming to gather data solely from physical experiments, especially in the dynamic field of AM. One way to circumvent that challenge is by utilising data analysis. In this thesis, Swee Leong statistically modelled the effect of SLM processing parameters on lattice structure properties using regression analysis and analysis of variance (ANOVA). It is shown that laser power and layer thickness are dominant factors in affecting the properties of the lattice structures that could correlate to better cellular attachment.

These favourable findings ascertained the feasibility of SLM TiTa as a 3D tool that can build a new biomaterial at the microstructure level, and contributed to the scientific knowledge of a novel in situ alloying process. The unique experimental methods and results, with the systematic analysis in this thesis, provided a valuable reference for researchers in the field of AM and 3D printing of metallic biomedical devices.

We have embarked on this journey 4 years ago when Swee Leong is an aspiring research student learning and eager to contribute to the scientific world. We are extremely proud that Swee Leong is now ready to be an impactful contributor to the scientific community and this thesis is the perfect gift to mark that milestone.

Singapore
September 2018

Dr. Wai Yee Yeong
Associate Professor
Dr. Florencia Edith Wiria
Research Scientist

Parts of this thesis have been published in the following journal articles:

1. S. L. Sing, F. E. Wiria & W. Y. Yeong (2018), *Selective laser melting of titanium alloy with 50 wt% tantalum: Effect of laser process parameters on parquality*, International Journal of Refractory Metals and Hard Materials 77, pp 120–127
2. S. L. Sing, F. E. Wiria & W. Y. Yeong (2018), *Selective laser melting of lattice structures: A statistical approach to manufacturability and mechanical behavior*, Robotics and Computer Integrated Manufacturing 49, pp 170–180
3. S. L. Sing,Y. Miao, F. E. Wiria & W. Y. Yeong (2016), *Manufacturability and mechanical testing considerations of metallic scaffolds fabricated using selective laser melting: A review*, Biomedical Science and Engineering 1(1), pp 18–24
4. S. L. Sing, W. Y. Yeong & F. E. Wiria (2016), *Selective laser melting of titanium alloy with 50 wt% tantalum: microstructure and mechanical properties*, Journal of Alloys and Compounds 660, pp 461–470
5. S. L. Sing, J. An, W. Y. Yeong & F. E. Wiria (2016), *Laser and electron-beam powder-bed additive manufacturing of metallic implants: a review on processes, materials and designs*, Journal of Orthopaedic Research 34(3), pp 369–385

Acknowledgements

First and foremost, I would like to express heartfelt gratitude and appreciation for my supervisors, Assistant Professor Wai Yee Yeong and Dr. Florencia Edith Wiria for their invaluable advice and patience throughout my candidature. Their dedication to the research has been a constant source of inspiration for me. Appreciation is also extended to Associate Professor Siaw Meng Chou and Dr. Bee Yen Tay for being members of my Thesis Advisory Committee.

I would like to thank Professor Chee Kai Chua, Associate Professor Kah Fai Leong and Dr. Jun Wei for the opportunity in pursuing my research in Singapore Centre for 3D Printing (SC3DP) and SIMTech-NTU Joint Lab (3D Additive Manufacturing). I would also like to acknowledge the research funding received from these institutions.

I am grateful for the companionship from fellow researchers in SC3DP who have begun this enriching journey with me, Dr. Alexander Zhonghong Liu, Dr. Danqing Zhang, Dr. Denise Loong Ee Loh and Dr. Chor Yen Yap. Their camaraderie and support during this study have been delightful. I would also like to express thanks to the other researchers who have helped me along the way.

This report would not have been possible without the assistance of the laboratory specialists from School of Mechanical and Aerospace Engineering, Nanyang Technological University and Singapore Institute of Manufacturing Technology.

Lastly, but certainly not the least, I would like to thank my family and friends for their constant encouragement and understanding.

Contents

1 Introduction to Additive Manufacturing for Orthopaedic Implants 1
1.1 Background 1
1.2 Motivations 3
1.3 Objectives 5
1.4 Scope 5
1.5 Organisation of Report 6
References 7

2 Concepts of Selective Laser Melting for Orthopaedic Implants 9
2.1 Orthopaedic Implants 9
2.2 Selective Laser Melting 12
2.3 Materials for Selective Laser Melting of Metallic Implants and Scaffolds 15
2.3.1 316L Stainless Steel 15
2.3.2 Cobalt Chromium 17
2.3.3 Titanium-6Aluminium-4Vanadium 17
2.3.4 New Materials 18
2.4 Design Consideration for Metallic Implants and Scaffolds Using Selective Laser Melting 20
2.4.1 Manufacturability of Metallic Scaffolds Using Selective Laser Melting 20
2.4.2 Biological Compatibility of Metallic Scaffolds Fabricated Using Selective Laser Melting 22
2.4.3 Mechanical Properties of Metallic Scaffolds Fabricated Using Selective Laser Melting 26
2.5 Summary 27
References 29

3 Formation of Titanium-Tantalum Alloy Using Selective Laser Melting 37
3.1 β Stabilisers for Titanium 37
3.2 Powder Preparation 37
3.3 Powder Characterization 38
3.4 Parameters Optimization for Selective Laser Melting 41
3.5 Summary 46
References 47

4 Characterisation of Selective Laser Melted Titanium-Tantalum Alloy 49
4.1 Metallographic Analysis of Titanium-Tantalum Parts 49
4.1.1 X-Ray Diffraction Phase Analysis 50
4.1.2 Microstructure 51
4.2 Mechanical Analysis of Titanium-Tantalum Parts 57
4.2.1 Tensile Properties 57
4.2.2 Micro-Hardness 61
4.3 Summary 61
References 62

5 Statistical Modelling of Selective Laser Melting of Cellular Lattice Structures 65
5.1 Forming Mechanisms of Lattice Structures Using Selective Laser Melting 65
5.2 Design of Cellular Lattice Structures 66
5.3 Design of Experiment for Regression Analysis 66
5.4 Properties Modelling of Cellular Lattice Structures 69
5.4.1 Metrological Characterisation 69
5.4.2 Mechanical Characterisation 75
5.5 Summary 83
References 84

6 Characterisation of Titanium-Tantalum Lattice Structures Fabricated by Selective Laser Melting 87
6.1 Design of Cellular Lattice Structures 87
6.2 Metrological Characterisation 87
6.3 Mechanical Characterisation 92
6.4 Biocompatibility of Titanium-Tantalum 93
6.5 Summary 95
References 95

7 Future Trends in Titanium-Tantalum Formed by Selective Laser Melting 97
7.1 In-Situ Alloying of Titanium-Tantalum using Selective Laser Melting 97
7.1.1 Formation of Bulk Titanium-Tantalum Using Selective Laser Melting 97
7.1.2 Formation of Lattice Structures Using Selective Laser Melting 98
7.2 Future Work 99
7.2.1 Development of Titanium-Tantalum Using Selective Laser Melting 99
7.2.2 Thermal Modelling of Selective Laser Melting 99
7.2.3 In Vitro and In Vivo Biocompatibility Tests for Scaffolds Fabricated by Selective Laser Melting 99

Abbreviations

3D	Three-dimensional
AM	Additive manufacturing
ANOVA	Analysis of variance
BCC	Body-centred cubic
CAD	Computer-aided design
cpTi	Commercially pure titanium
CVD	Chemical vapour deposition
DOE	Design of experiment
EBSD	Electron backscattered diffraction
EDM	Electrical discharge machining
EDS	Energy-dispersive X-ray spectroscopy
FE	Finite element
FESEM	Field emission scanning electron microscopy
HCP	Hexagonal close packed
ICP-AES	Inductive coupled plasma atomic emission spectroscopy
OM	Optical microscopy
SLM	Selective laser melting
Ti6Al4V	Titanium-6aluminium-4vanadium
TiTa	Titanium-tantalum
XRD	X-ray diffraction

Symbols

a_i	Correction constant coefficient
a_j	Coefficient for linear effect
a_{ii}	Coefficient for quadratic effect
a_{ij}	Coefficient for interaction effect
C_E	Gibson–Ashby model parameter for elastic constant
C_Y	Gibson–Ashby model parameter for yield strength
D	Diagonal strut dimension
E	Theoretical elastic constant
E_0	Elastic constant of the fully dense material
E_d	Energy density
F	Test statistic
G	Temperature gradient in liquid phase
H	Hatch spacing
H_S	Horizontal strut dimension in yz-plane
H_T	Horizontal strut dimension in xy-plane
k_E	Gibson–Ashby model constant for elastic constant
k_Y	Gibson–Ashby model constant for yield strength
L	Layer thickness
Mo_{eq}	Molybdenum equivalence
m_{air}	Mass of specimen in dry weighing
m_{water}	Mass of specimen fully submerged in deionised water
p	Probability
P	Laser power
P_r	Porosity of lattice structure
R	Solidification rate
R^2	Data fit
V	Vertical strut dimension
x_i	Independent variable
x_j	Independent variable
Y_S	Theoretical yield strength

Y_0	Yield strength of fully dense material
ε	Random error
ρ_{abs}	Absolute density of specimen
$\rho_{relative}$	Relative density of specimen
$\rho_{theoretical}$	Theoretical density of specimen
ρ_{water}	Density of deionised water

Chapter 1
Introduction to Additive Manufacturing for Orthopaedic Implants

This chapter provides the background, motivations, objectives and scope of the thesis. The background discusses the capability of selective laser melting (SLM) in manufacturing of metallic functional parts directly. As a laser powder bed fusion (L-PBF) technology, it also has the potential for forming alloys from mixed powders directly which motivates this work. The main objective and scope of this work are to develop and characterise novel titanium alloy for biomedical applications which is further elaborated in this chapter. The thesis organisation is also presented as a brief overview of the content of this thesis.

1.1 Background

Additive manufacturing (AM), also commonly known as 3D printing, is a group of processes that make objects by joining materials together, usually layer-by-layer, based on data from three-dimensional (3D) models. This is in contrast to conventional manufacturing methodologies that are subtractive [1]. A schematic of typical AM process chain is shown in Fig. 1.1.

These technologies have the potential to impact the designs and applications in multiple fields such as automotive [2], aerospace [3], biomedical [4–8], electronics [9] and even common goods such as jewellery and fashion [10]. In particular, AM techniques have been used in multiple studies on tissue engineering [4, 11–15]. With the advancement in AM techniques and more materials becoming available for them, the functionality of AM has been extended to the field of orthopaedic implants and scaffolds.

Orthopaedic implants as artificial bones, or scaffolds are mostly used for structural reinforcement when inserted inside the body. They include both temporary implants, such as screws and plates, and permanent implants that are used to replace body parts such as hip and knee directly [16]. Permanent implants put more emphasis on toughness, strength, and tribology as well as abrasion resistance between artificial joints.

S. L. Sing, *Selective Laser Melting of Novel Titanium-Tantalum Alloy as Orthopaedic Biomaterial*, Springer Theses, https://doi.org/10.1007/978-981-13-2724-7_1

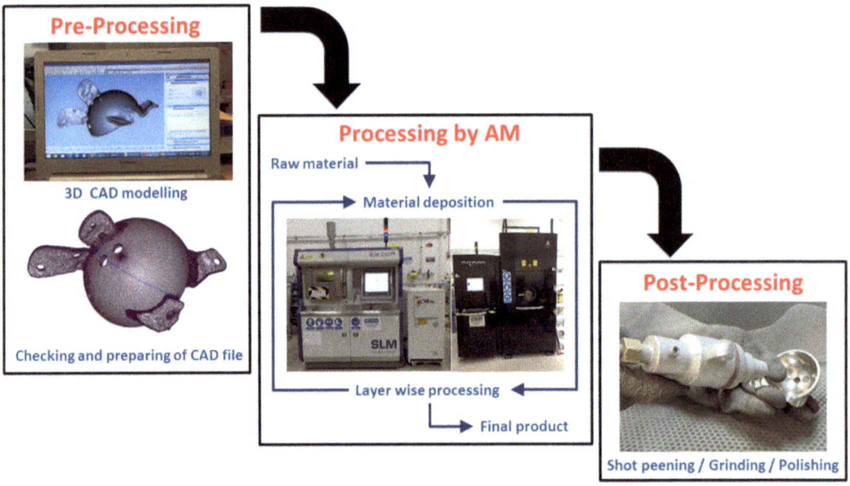

Fig. 1.1 AM process chain

For patients with extensive bone loss or deformities, commercial implants often do not provide an acceptable clinical solution presently. The current implants are usually made of one material, which is essentially uniform in composition and structure in the longitudinal direction [17, 18]. This leads to identical properties, such as strength and biocompatibility, throughout the implant. However, a uniform structure with single composition cannot satisfy all the requirements needed for implants [19]. For specific bone tissues, such as long bones, bone porosity varies from the outer to the inner section in order to achieve normal bone function. In this case, the bone's mechanical strength decreases gradually from the outer to inner regions. Hence, the bone can be regarded as a functionally graded structure [12]. To solve this problem, current implants may have a coating layer of different material or incorporate a porous outer layer that are manufactured separately and fitted to the implants. The conventional methods of manufacturing implants therefore require multiple steps. Furthermore, current implants are manufactured in sizes that are not customised to the patients. As a result, longer duration is needed in order to do the implant fitting to the patient during the surgery [20, 21]. Customised implants have the potential to reduce surgery, rehabilitation and recovery times, improve implant fixation and reduce the likelihood of revision surgery.

With AM technologies, such as SLM, that are capable of manufacturing functional metallic parts directly, next generation orthopaedic implants can be manufactured in a single step [22]. The implants can be designed with improved matching of mechanical strength and optimised for osteo-integration. Furthermore, they can be customised for specific patients. These can be achieved as SLM has more design freedom as compared to conventional manufacturing techniques, thus allowing it to build complex geometries without significant increase in building time. In addition, SLM requires no tooling or moulds, hence enabling the fabrication of several patients'

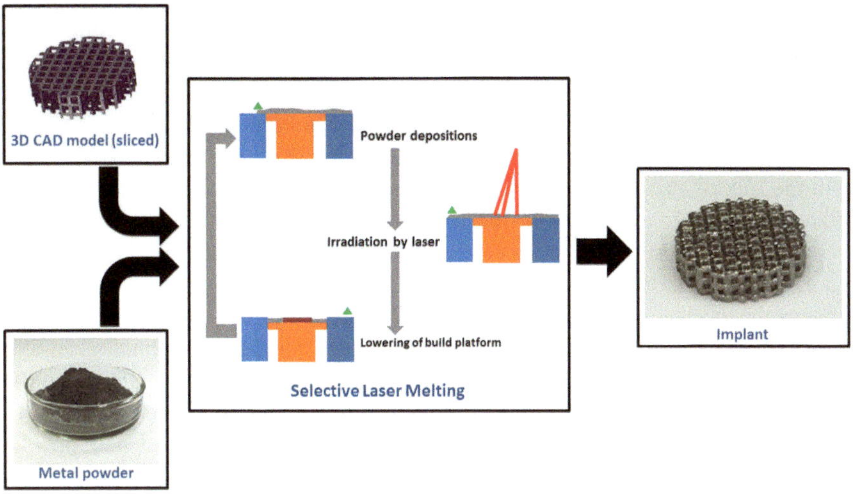

Fig. 1.2 Key steps in SLM

implants in the same batch. SLM is able to provide lesser design constraints to product developers and significantly lower the customization cost [1].

According to ISO/ASTM 52900:2015 (Standard Terminology for Additive Manufacturing—General Principles—Terminology), SLM is classified as a powder bed fusion process where thermal energy from a laser selectively fuse regions of a powder bed. Powder bed fusion processes use an energy source to melt and fuse selective regions of powder according to sliced computer aided design (CAD) data. When the selective melting of one layer is completed, the building platform is lowered by a predetermined distance (usually 20–100 μm) and a next layer of powder is deposited on the building platform. The process is then repeated with successive layers of powder until the required part is completely built [1]. The exclusion of sacrificial binders in the process enables near-full density parts to be built. This gives it a significant advantage over binder-based processes in direct part manufacturing. An overview of key steps within SLM is shown in Fig. 1.2.

1.2 Motivations

As discussed, the literatures reviewed show the limitations of conventional methods in manufacturing metallic implants. This means that there remains a lot of work that can be done to improve current implants that are in the market. The key areas that can lead to improvement in implants are identified to be material development and design improvement.

Human bones have wide range of elastic constants, for example, from 1.0 to 25.0 GPa [23, 24]. However, the elastic constant for biocompatible metals is often much higher. This modulus mismatch can result in an adverse effect called "stress shielding" [25]. Stress shielding induces an undesirable stress distribution at the bone-implant interface, resulting in slower bone healing [8, 25] as the bone remodels itself from the lack of stress stimulant. Clinical investigations indicate that the mismatch will result in insufficient load transfer from artificial implants to neighbouring bones, resulting in bone resorption and potential loosening of the implant [26]. Thus, there is a need to develop materials of reduced modulus to avoid mismatch in modulus between the implant and adjacent bones. Material development can lead to materials with high strength to modulus ratio. Using powder bed fusion technologies, it is now possible to obtain alloys that were difficult to mix previously. The capability of SLM to process powder mixtures has opened up exciting new material research opportunities. Several works have been reported on several types of new powder mixture processed by SLM [27–31].

Tantalum is an excellent choice for alloying with titanium for biomedical applications due to its high biocompatibility, corrosion resistance and good mechanical properties. Furthermore, titanium-tantalum (TiTa) alloys are promising materials for such applications because of high strength to density ratio [32]. Alloying elements in titanium can be classified into three groups: (1) α phase stabiliser, (2) β phase stabiliser or (3) elements that have no observable effect on the phase [31]. Depending on the specific application of the materials, the different phases of titanium provide a wide array of properties. Tantalum is a β stabilizing element for titanium alloy. β titanium alloys display superior properties with lower modulus compared to the commonly used alloys in the biomedical field, such as stainless steels and cobalt-chromium alloys [28] and Ti6Al4V which is an ($\alpha + \beta$) titanium alloy [33].

Despite the advantages, TiTa alloys are still not widely adopted in applications. The main reason is the difficulty in combining these two metals as they have great difference in density and melting point [33]. Tantalum has a density of 16.6 g/cm^3 which is about four times the density of commercially pure titanium (cpTi). This could lead to inhomogeneity during the alloy formation as the large difference in density can lead to segregation of elements in the alloys. SLM provides the opportunity in creating a homogenous alloy from titanium and tantalum.

In addition to manufacturing parts with complex geometries, SLM can fabricate metallic lightweight structures, such as cellular lattice structures. These structures provide advantages such as high strength to weight ratio and high performance which makes them suitable for high value industrial applications such as medical implants [34]. Improved designs of the implants can be achieved through the introduction of porosity by creating lattice structures. With lattice structures, it is also possible to minimize the stress shielding via the reduction of elastic modulus of the implants. Porous lattice structures also improve osteo-integration. By using SLM, there is greater freedom in designing and inducing porosity on the implants and the porosity of such structures can be controlled with better precision. The design freedom and reproducibility are importance features for implants, especially if there is a need for performance simulation and outcome prediction [35, 36].

1.3 Objectives

The main objective of the research is to optimise the SLM process for manufacturing of implants. Through this research, the understanding of SLM process will improve the capability of manufacturing process of implants. The objectives of this research are:

1. To develop a novel biocompatible titanium alloy that was previously difficult to obtain by using SLM. The forming mechanism and microstructure relationships to the thermal phenomena experienced during the SLM process will be studied.
2. To obtain homogenous TiTa alloy from SLM. The mechanical properties and biocompatibility of this alloy will be characterised and benchmarked against more commonly used titanium alloy such as Ti6Al4V and cpTi. The relationship between microstructure and mechanical properties of these materials will also be studied in depth.
3. To understand the process of lattice fabrication using SLM. The limits and feasibility of SLM in lattice structure fabrication will be ascertained. Relationship between the process parameters and actual properties of lattice structures will be derived.

1.4 Scope

The scope of this research is as follows:

1. X-ray diffraction (XRD), field emission scanning electron microscopy (FESEM) and electron backscattered diffraction (EBSD) will be carried out to investigate the properties of titanium alloys fabricated using SLM.
2. The metallic scaffolds will be characterised based on dimensions, porosity and compression properties to understand the basic phenomena that occur in SLM for metallic scaffolds fabrication. Statistical modelling will be done to understand the relationships.
3. Standard tests obtained from ASTM and ISO will be carried out to mechanically characterise the SLM fabricated samples.

The approach to achieve the objectives and scope of the research is summarised in Fig. 1.3.

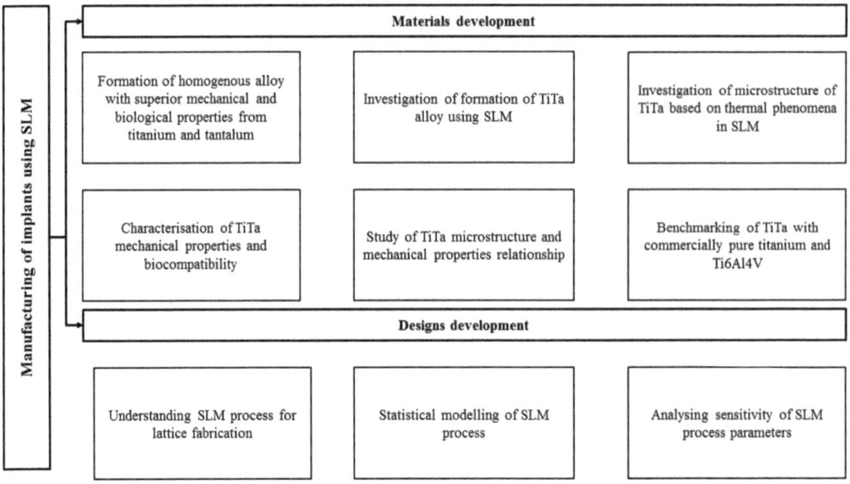

Fig. 1.3 Research overview

1.5 Organisation of Report

This thesis consists of seven chapters.

This thesis begins with the introduction, describing the background of the research and stating the objectives and scope in Chapter 1. It is then followed by literature review to present related theory, methods and findings of past research in Chap. 2.

Chapter 3 gives details of the novel TiTa alloy formation using SLM. Powder preparation method, powder characterisation and criteria for SLM optimisation in material development are highlighted.

The microstructures and mechanical properties of the TiTa alloy are characterised in Chap. 4. The science behind the alloy formation is also explained using thermal phenomena experienced during SLM. The effect of SLM processing parameters on the fabrication of lattice structures is statistically modelled in Chap. 5 the relationships between the processing parameters to lattice structures dimensions, porosity and compression behaviour are then explored.

Building from the understanding derived in Chaps. 5 and 6 details the studies on the mechanical and in vitro test on the SLM fabricated scaffolds. The compression behaviour and biocompatibility of TiTa scaffolds are benchmarked against cpTi and Ti6Al4V scaffolds.

Finally, in Chap. 7, conclusions and possible future work on this research are presented.

References

1. C.K. Chua, K.F. Leong, *3D Printing and Additive Manufacturing: Principles and Applications*, 5th edn. (World Scientific Publishing Co. Pte. Ltd, Singapore, 2017)
2. Z.H. Liu, D.Q. Zhang, S.L. Sing, C.K. Chua, L.E. Loh, Interfacial characterization of SLM parts in multi-material processing: metallurgical diffusion between 316L stainless steel and C18400 copper alloy. Mater. Charact. **94**, 116–125 (2014)
3. C. Yan, L. Hao, A. Hussein, P. Young, D. Raymont, Advanced lightweight 316L stainless steel cellular lattice structures fabricated via selective laser melting. Mater. Des. **55**, 533–541 (2014)
4. W.Y. Yeong, N. Sudarmadji, H.Y. Yu, C.K. Chua, K.F. Leong, S.S. Venkatraman, Y.C. Boey, L.P. Tan, Porous polycaprolactone scaffold for cardiac tissue engineering fabricated by selective laser sintering. Acta Biomater. **6**, 2028–2034 (2010)
5. W.Y. Yeong, C.K. Chua, A quality management framework for implementing additive manufacturing of medical devices. Virtual and Phys. Prototyp. **8**, 193–199 (2013)
6. M.W. Naing, C.K. Chua, K.F. Leong, Y. Wang, Fabrication of customised scaffolds using computer-aided design and rapid prototyping techniques. Rapid Prototyp. J. **11**, 249–259 (2005)
7. A. Barbas, A.-S. Bonnet, P. Lipinski, R. Pesci, G. Dubois, Development and mechanical characterization of porous titanium bone substitutes. J. Mech. Behav. Biomed. Mater. **9**, 34–44 (2012)
8. P. Heinl, L. Muller, C. Korner, R.F. Singer, F.A. Muller, Cellular Ti-6Al-4 V structures with interconnected macro porosity for bone implants fabricated by selective electron beam melting. Acta Biomater. **4**, 1536–1544 (2008)
9. S.L. Sing, L.P. Lam, D.Q. Zhang, Z.H. Liu, C.K. Chua, Interfacial characterization of SLM parts in multi-material processing: Intermetallic phase formation between AlSi10 Mg and C18400 copper alloy. Mater. Charact. **107**, 220–227 (2015)
10. Y.L. Yap, W.Y. Yeong, Additive manufacturing of fashion and jewellery products: a mini review. Virtual Phys. Prototyp. **9**, 195–201 (2014)
11. C.K. Chua, W.Y. Yeong, *Bioprinting: Principles and Applications* (World Scientific Publishing Co Pte. Ltd, Singapore, 2014)
12. N. Sudarmadji, J.Y. Tan, K.F. Leong, C.K. Chua, Y.T. Loh, Investigation of the mechanical properties and porosity relationships in selective laser-sintered polyhedral for functionally graded scaffolds. Acta Biomater. **7**, 530–537 (2011)
13. F.E. Wiria, K.F. Leong, C.K. Chua, Y. Liu, Poly-e-caprolactone/hydroxyapatite for tissue engineering scaffold fabrication via selective laser sintering. Acta Biomater. **3**, 1–12 (2007)
14. S.F. Yang, K.F. Leong, Z.H. Du, C.K. Chua, The design of scaffolds for use in tissue engineering. Part II. Rapid prototyping techniques. Tissue Eng. **8**, 1–11 (2002)
15. W.Y. Yeong, C.K. Chua, K.F. Leong, M. Chandrasekaran, Rapid prototyping in tissue engineering: challenges and potential. Trends Biotechnol. **22**, 643–652 (2004)
16. G. Manivasagam, D. Dhinasekaran, A. Rajamanickam, Biomedical implants: corrosion and its prevention—a review. Recent Pat. Corros. Science **2**, 40–54 (2010)
17. F. Watari, A. Yokoyama, M. Omori, T. Hirai, H. Kondo, M. Uo, T. Kawasaki, Biocompatibility of materials and development to functionally graded implant for bio-medical application. Compos. Sci. Technol. **64**, 893–908 (2004)
18. P. Muller, P. Mognol, J.-Y. Hascoet, Modeling and control of a direct laser powder deposition process for functionally graded materials (FGM) parts manufacturing. J. Mater. Process. Technol. **213**, 685–692 (2013)
19. F. Watari, A. Yokoyama, F. Saso, M. Uo, T. Kawasaki, Fabrication and properties of functionally graded dental implant. Composit. Part B: Eng. **28**, 5–11 (1996)
20. A.L. Jardini, M.A. Larosa, C.A. de Carvalho Zavaglia, L.F. Bernardes, C.S. Lambert, P. Kharmandayan, D. Calderoni, R.M. Filho, Customised titanium implant fabricated in additive manufacturing for craniomaxillofacial surgery. Virtual Phys. Prototyp. **9**, 115–125 (2014)
21. A.L. Jardini, M.A. Larosaemail, R.M. Filho, C.A.d.C. Zavaglia, L.F. Bernardes, C.S. Lambert, D.R. Calderoni, P. Kharmandayan, Cranial reconstruction: 3D biomodel and custom-built implant created using additive manufacturing. J. Cranio-Maxillo-Fac. Surg. (2014)

22. R. Wauthle, S.M. Ahmadi, S. Amin Yavari, M. Mulier, A.A. Zadpoor, H. Weinans, J. Van Humbeeck, J.-P. Kruth, J. Schrooten, Revival of pure titanium for dynamically loaded porous implants using additive manufacturing. Mater. Sci. Eng., C **54**, 94–100 (2015)
23. P.K.Z. Zysset, X.E. Guo, C.E. Hoffler, K.E. Moore, S.A. Goldstein, Elastic modulus and hardness of cortical and trabecular bone lamellae measured by nanoindentation in the human femur. J. Biomech. **32**, 1005–1012 (1999)
24. B. Piotrowski, A.A. Baptista, E. Patoor, P. Bravetti, A. Eberhardt, P. Laheurte, Interaction of bone-dental implant with new ultra low modulus alloy using a numerical approach. Mater. Sci. Eng., C **38**, 151–160 (2014)
25. T. Traini, C. Mangano, R.L. Sammons, F. Mangano, A. Macchi, A. Piattelli, Direct laser metal sintering as a new approach to fabrication of an isoelastic functionally graded material for manufacture of porous titanium dental implants. Dental Mater. **24**, 1525–1533 (2008)
26. P. Laheurte, F. Prima, A. Eberhardt, T. Gloriant, M. Wary, E. Patoor, Mechanical properties of low modulus β titanium alloys designed from the electronic approach. J. Mech. Behav. Biomed. Mater. **3**, 565–573 (2010)
27. H. Tan, J. Chen, F. Zhang, X. Lin, W. Huang, Microstructure and mechanical properties of laser solid formed Ti-6Al-4 V from blended elemental powders. Rare Metal Mater. Eng. **38**, 574–578 (2009)
28. D. Gu, Y.-C. Hagedorn, W. Meiners, K. Wissenbach, R. Poprawe, Selective laser melting of in-situ TiC/Ti5Si3 composites with novel reinforcement architecture and elevated performance. Surf. Coat. Technol. **205**, 3285–3292 (2011)
29. D. Gu, W. Meiners, Microstructure characteristics and formation mechanisms of in situ WC cemented carbide based hardmetals prepared by selective laser melting. Mater. Sci. Eng., A **527**, 7585–7592 (2010)
30. D. Gu, G. Meng, C. Li, W. Meiners, R. Poprawe, Selective laser melting of TiC/Ti bulk nanocomposites: influence of nanoscale reinforcement. Scr. Mater. **67**, 185–188 (2012)
31. B. Vrancken, L. Thijs, J.-P. Kruth, J. Van Humbeeck, Microstructure and mechanical properties of a novel β titanium metallic composite by selective laser melting. Acta Mater. **68**, 150–158 (2014)
32. K.A. de Souza, A. Robin, Preparation and characterization of Ti-Ta alloys for application in corrosive media. Mater. Lett. **57**, 3010–3016 (2003)
33. A. Morita, H. Fukui, H. Tadano, S. Hayashi, J. Hasegawa, M. Niinomi, Alloying titanium and tantalum by cold crucible levitation melting (CCLM) furnace. Mater. Sci. Eng., A **280**, 208–213 (2000)
34. C. Yan, L. Hao, A. Hussein, D. Raymont, Evaluations of cellular lattice structures manufactured using selective laser melting. Int. J. Mach. Tools Manuf. **62**, 32–38 (2012)
35. S.M. Ahmadi, G. Campoli, S. Amin Yavari, B. Sajadi, R. Wauthle, J. Schrooten, H. Wienans, A.A. Zadpoor, Mechanical behavior of regular open-cell porous biomaterials made of diamond lattice unit cells. J. Mech. Behav. Biomed. Mater. **34**, 106–115 (2014)
36. G. Campoli, M.S. Borleffs, S. Amin Yavari, R. Wauthle, H. Weinans, A.A. Zadpoor, Mechanical properties of open-cell metallic biomaterials manufactured using additive manufacturing. Mater. Des. **49**, 957–965 (2013)

Chapter 2
Concepts of Selective Laser Melting for Orthopaedic Implants

This chapter introduces biological implants and the conventional methods in manufacturing of biological implants. The state-of-the-art technology of selective laser melting (SLM) is then presented. It provides a comprehensive review on the process, materials, properties and designs with focus on biomedical applications. Challenges and potential of SLM in the biomedical field are also highlighted.

2.1 Orthopaedic Implants

In general, there are "orthopaedic implants" as artificial bones for medical use and "dental implants" as artificial teeth for dental use [1, 2]. The implants specific properties are slightly different depending on their use. Orthopaedic implants and scaffolds are mostly used as artificial bone to provide structural reinforcement when inserted into the body. They can include both permanent and temporary implants [3]. While temporary implants such as screws and plates are removed from the body, permanent implants, such as hip and knee, need to have a longer durability in the body. Therefore, permanent implants focus more on strength, toughness, abrasion resistance in artificial joints as well as tribology. SLM has been utilized to fabricate orthopaedic implants such as replacements for zygomatic bone (forming the prominent part of the cheek and the outer side of the eye socket) [4] and finger [5]. Dental implants are usually much smaller in size and used to reconstruct the masticatory function, when the tooth root is completely lost or extracted. The major classifications of dental implants are endosseous implants, which are placed into the bone, and the subperiosteal implants which are placed on top of the bone. The major reasons for corrosion in metallic dental implants are temperature, quantity and quality of saliva which has pH of between 5.2 and 7.8, the physical and chemical properties of food and liquids and oral health conditions [3]. Depending on the requirement of the implants, some of them may require post-processing, such as mechanical polishing, grinding or sandblasting.

S. L. Sing, *Selective Laser Melting of Novel Titanium-Tantalum Alloy as Orthopaedic Biomaterial*, Springer Theses, https://doi.org/10.1007/978-981-13-2724-7_2

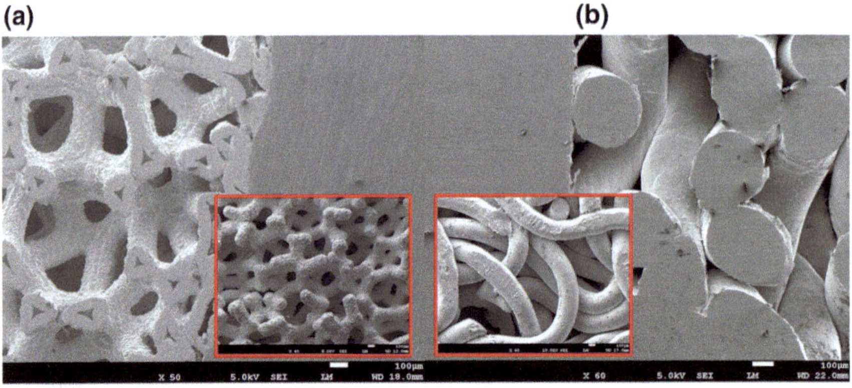

Fig. 2.1 Porous surfaces of implants **a** acetabular cup **b** tibial tray with zoomed in images of uncontrolled porosity

The dental implant is used in a manner that results in the implant penetrating the jaw bone and leaving part of it exposed outside of the bone. Hence, the function is different at the inside and outside of the bone, and at their boundary. For example, bone affinity is important inside the jaw bone and at the outside, in the oral cavity, sufficient strength is needed [1]. Natural bone is a composite in which nano-sized bone minerals are deposited on organic collagen fibres, which are woven into complex three-dimensional (3D) structures [6–8]. Current metallic implants are essentially neutral in vivo, remaining as permanent fixtures. However, there are some cases where the implants must be removed by a second surgical procedure after the tissue has healed sufficiently [9].

Typically implants are fixed to the bone in several ways, including the use of bone cement, mechanical fixation devices such as screws, interference fits and activated surfaces that result in bone apposition (both in-growth or on-growth) [10]. Bone in-growth refers to the bone formation within an irregular (beads, wire mesh, casting voids, cut grooves) surface of an implant while on-growth refers to the bone to implant contact. All these have been successful in applications and are still being used, however, bone in-growth has gained popularity recently by the use of porous biomaterials [10]. An added advantage of porous biomaterials is they are inherently less stiff than non-porous counterparts, and therefore, the mismatch of implant-bone stiffness is less pronounced, thereby reducing the possibility of stress shielding.

Implants are conventionally produced by machining metal rods, casting or forging. These are usually followed by modification of the surface by applying different surface treatments, coatings or another porous part to improve stability and enhance osteo-integration [11]. These porous layers are usually manufactured by conventional techniques, such as furnace sintering, plasma spraying, wax casting and vapour deposition [12–14]. Due to the constraints of the manufacturing process, they often have random porosity that cannot be controlled. Some examples of porous surface on implants in the market currently are shown in Fig. 2.1.

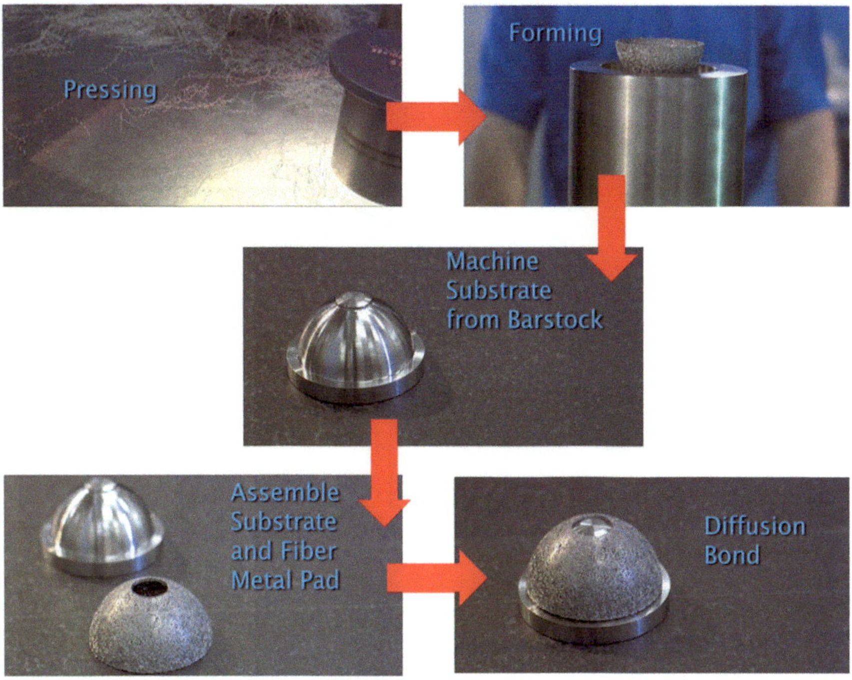

Fig. 2.2 Conventional multiple steps manufacturing process of implants. (Adapted from Zimmer Biomet Holdings, Inc)

Multiple steps are therefore needed for conventional manufacturing of a complete implant, as shown in the manufacturing process in Fig. 2.2.

For the porous component of implants, metal foams are already in used for structures employed for biomedical applications, however, the resulting stochastic geometry brings about unpredictable localized mechanical properties [15]. A variety of methods to manufacture porous systems have been developed, such as sintered beads, fibre mesh and thermal spray processes. However, they have disadvantages such as compromised microstructure, incapability to achieve interconnected pores at optimal porosity level and lack of porosity controllability [16]. A recently developed process, chemical vapour deposition (CVD) addresses this issue. However, it is a multi-stage process that is complex and incapable of producing parts directly with minimal downstream operations [16, 17]. Freeze casting technique enables a porous material with a compressive strength close to that of bone to be obtained, but its disadvantage is that the pore size cannot exceed 300 μm. Larger pore sizes can be obtained with the space holder technique but the mechanical properties obtained are lower than those of human bone. Non-homogenous porous metal samples are also fabricated using the space holder method [18]. Hence, the control of the elastic properties anisotropy is difficult for these techniques [19, 20].

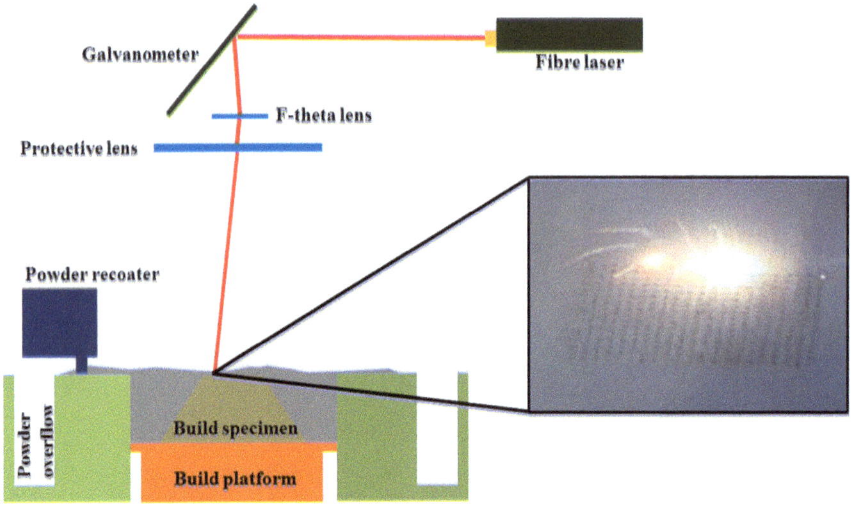

Fig. 2.3 Schematic of SLM system

2.2 Selective Laser Melting

SLM uses a fibre laser as the energy source [21], hence, it is also known as laser powder bed fusion (L-PBF). The whole process is carried out in an inert gas filled or vacuum chamber which ensures higher purity by minimizing the oxygen in the environment and reducing the risk of hydrogen pick up. The schematic of the SLM system is shown in Fig. 2.3.

Depending on the laser module installed in the SLM system, the laser can operate up to 1 kW [22]. The beam focus is controlled by the galvanometer and the movement of the beam on the build platform is controlled by F-theta lens. For building a part, powder layers between 20 and 100 μm thickness are spread over the build platform. The powder is carried and spread by the powder recoater across the build platform. The build platform can be preheated up to 200 °C. The selective melting of the powder layer based on the geometry defined by the sliced CAD file is done by the laser. In SLM, every layer of a part is built in two steps. Typically, the outer boundary of the part is built first which is referred as contouring and this can be optional. The powder within the contour is melted subsequently to complete one layer. This process continues until the desired 3D part is fully completed [23].

Orthopaedic implants and scaffolds are traditionally manufactured by machining, casting and forging. While these methods are the industry standard that have been certified and proven to safely manufacture the implants and scaffolds, AM has slowly been influencing the manufacturing dynamic of customised implants and scaffolds.

AM techniques, including SLM, have numerous advantages and are actively considered for manufacturing of orthopaedic implants and scaffolds. These advantages include:

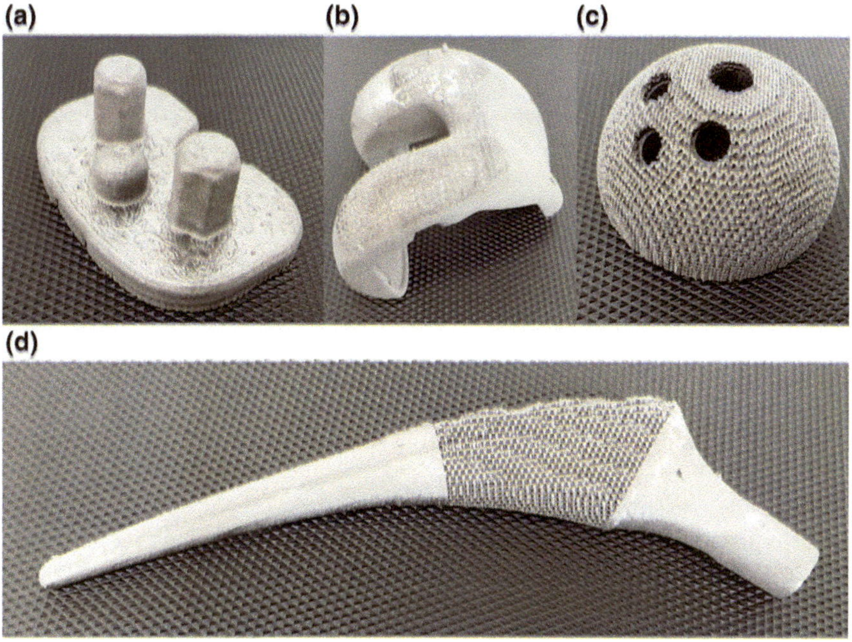

Fig. 2.4 Hip and knee implants fabricated by SLM in-house **a** tibial tray **b** femoral component **c** acetabular cup **d** hip stem

- Fabrication of complex products with novel shapes and hollow structures that are not otherwise feasible
- Forming functionally graded materials or structures
- Efficient approach that reduces production costs and speed up time-to-market for high-value components
- Excellent material properties with almost no porosity
- Possible to combine different materials, significantly reduced material waste
- Elimination of expensive tooling and moulds.

Leveraging on these advantages, numerous studies have been carried out to prove the feasibility of SLM in producing implants as shown in Fig. 2.4.

The primary objective in SLM is to obtain defect free parts with near full density. Many factors affect the quality of the fabricated parts and they are tabulated in Table 2.1. With so many factors in consideration, appropriate control of the related factors is needed to fabricate products with high quality.

Several laser parameters (such as laser wavelength, laser working mode and spot size) cannot be varied and are determined by the SLM machine. In addition, several material parameters, such as thermal properties (for example, thermal conductivity), are fixed for a given powder. These define the boundary conditions of SLM process.

Table 2.1 Factors involved in SLM. (Adapted from [24])

Materials parameters	Laser parameters	Scan parameters	Environment parameters
Powder composition	Laser scan mode (continuous/pulsed)	Scanning speed	Preheating temperature
Relative density (when deposited)	Laser wavelength	Scan distance	Chamber pressure
Particle size distribution	Laser power	Hatch spacing	Gas environment
Shape	Frequency	Layer thickness	Gas flow rate
Thermal properties	Laser pulse width	Scan pattern	Part placement
Flow properties	Offset	Scan sectors	Part orientation
	Laser spot size	Scaling factors	

The key processing parameters involved in SLM are laser power, scanning speed, hatch spacing and layer thickness [13, 23]. These factors determine the energy supplied by the laser beam to a volumetric unit of powder material, defined as energy density E_d.

$$E_d = \frac{P}{S.H.L} \tag{2.1}$$

where P is the laser power, S is the scanning speed, H is the hatch spacing and L is the layer thickness. The energy density influences the quality of parts fabricated by SLM by affecting the physical densification. Hence, the energy density determines the properties of the SLM produced parts. During SLM, the laser beam moves across the powder bed at a constant rate, defined by the scanning speed. The scanning speed controls the time needed for fabrication. Hence, higher scanning speed is desired if short production time is needed. However, the maximum scanning speed that can be used is limited by the maximum power of the laser in the SLM machine as high scanning speed with low laser power will results in incomplete melting of the material. Layer thickness determines the amount of energy required to completely melt and consolidate a powder layer. Layer thickness is very important as good connectivity between layers can only be possible when previously processed layers are partially remelted too. The production time is reduced if larger layer thickness is applied. However, higher energy input is also required to melt thicker layers completely. Larger layer thickness may cause the increase in surface roughness and reduction in dimensional accuracy. It has been reported that hatch spacing influences the overlap and bonding of adjacent melt tracks, therefore, affecting the porosities and surface roughness of SLM fabricated parts [23, 25].

The scan strategy, also known as scan pattern, can be designed in many ways. The scan strategy typically consists of straight and parallel scan tracks. The direction of these scan strategies or the single scan tracks can be changed within a single layer or

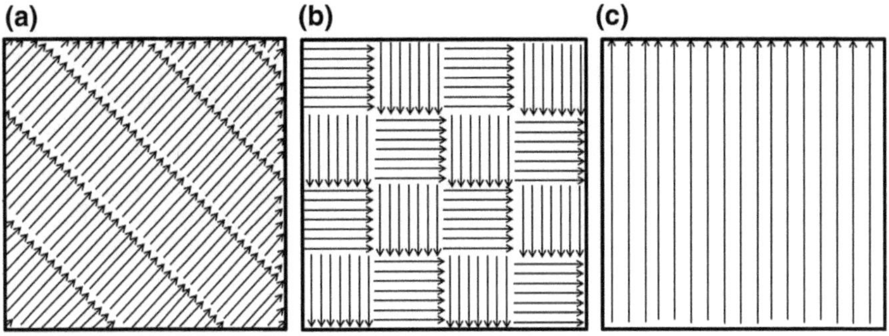

Fig. 2.5 SLM scan strategy **a** stripes **b** chessboard **c** meanders

between consecutive layers. The design of these scan strategies influences the quality of SLM parts. Some examples of scan strategy are shown in Fig. 2.5.

During SLM, thermal energy causes the melting and consolidation of the powder in which the temperature can rise even above the evaporation point. Hence, the material properties are important in melt pool formation. The powder morphology, such as particle size and shape, plays a key role in determining the quality of the products. Powder morphology determines the extent of how the powder particles are packed together during the layer deposition. Therefore, it can affect the layer thickness and relative density of the deposited layer which in turn, has effect on the quality of the produced parts. Ideally, the powder particles should be spherical for good flowability [26].

2.3 Materials for Selective Laser Melting of Metallic Implants and Scaffolds

The widely known biocompatible materials established for SLM are 316L stainless steel, cobalt chromium alloys and Ti6Al4V. Recent researches focus on expanding the material library for SLM [27–31], process simulations [32, 33], as well as characterizing and improving the mechanical properties of established materials [34–36]. In this section, the key findings of research for these materials will be highlighted.

2.3.1 316L Stainless Steel

316L stainless steel is low cost, biocompatible and easily available, making it a suitable material choice in the medical industry for metallic implants. Coupled with SLM, 316L stainless steel is well suited for orthopaedic applications as implants or prostheses that can be individualized with minimal customization costs.

Table 2.2 Tensile properties and micro-hardness of 316L stainless steel by SLM and forging

Properties	SLM	Forging
Ultimate tensile strength	480–760 MPa [13, 26, 46]	450–818 MPa [47, 48]
Yield strength	350–640 MPa [46, 49]	150–230 MPa [47, 50]
Elongation	10–30% [13, 49]	50–62% [47, 48]
Micro-hardness	220–279 HV [13, 26]	133–140 HV [47]

Yang et al. studied the optimization of building accuracy and density of orthodontic products using 316L stainless steel. They were able to achieve the required mechanical properties and surface quality using a self-developed SLM machine [37]. Li et al. investigated the feasibility of making SLM 316L stainless steel parts with graded porosity where the dense portion is designed for strength and the porous part is designed to enhance tissue growth in biocompatible implants [38]. Bibb et al. and Kruth et al. produced SLM frameworks for denture using the same material in different studies [39, 40]. Bibb et al. also presented four case studies on surgical guides in different maxillofacial (jaw and face) surgeries [41]. Wehmoller et al. reported body implants of cortical bone, mandibular canal segment (near the lower jaw) and support structures or tubular bone made from SLM 316L stainless steel [42].

Mechanical Properties

Relative Density

The relative density of a part is often used as a quality indicator for the SLM produced part. Relative density is the ratio of the density of the SLM fabricated part to the theoretical density of the bulk material.

Tolosa et al. was able to achieve a relative density of 99.90% [43] and Yasa et al. achieved 99.95% relative density with laser remelting [44] for 316L stainless steel. While remelting process is able to achieve near dense relative density, laser remelting increases the power consumption and fabrication process time as each layer is scanned twice.

Strength and Hardness

316L stainless steel is often used in many applications due to its strength. SLM produced steel components are stronger and less malleable as compared to the same material produced by forging [45]. Refined microstructure, because of rapid cooling in SLM, improves the tensile strength but reduces the ductility.

The mechanical properties of 316L stainless steel parts produced by SLM and forging are shown in Table 2.2.

A fine cellular dendritic structure is characteristic of SLM produced 316L stainless steel parts which are due to the rapid solidification during the SLM process. This results in the higher strength obtained as compared to forged 316L stainless steel parts which experience slower cooling rates.

Table 2.3 Tensile strengths and micro-hardness of SLM and casted CoCr

Properties	SLM	Casting
Ultimate tensile strength	951–1308 MPa [56–58]	591–759 MPa [56, 57]
Yield strength	562–884 MPa [56–58]	296–568 MPa [56, 57]
Elongation	10.2–16.4% [56–58]	8.0–10.7% [56, 57]
Micro-hardness	458.3–482.0 HV [54, 59]	324.0–384.8 HV [59, 60]

Surface Roughness

Post-processing such as sand blasting, shot peening or manual grinding is often needed to achieve a smooth and shiny surface for SLM parts. However, Delgado et al. were able to achieve surface roughness of 5.82 μm for 316L stainless steel without any post-processing [51]. Kruth's group was able to achieve surface roughness as low as 2 μm for 316L stainless steel without post-processing by using laser remelting [44].

2.3.2 *Cobalt Chromium*

Cobalt chromium alloys produced by SLM have been studied by various groups for implant applications. Oyague et al. and Kim et al. evaluated the fit of dental prostheses produced by SLM separately. Different conclusions about the suitability of SLM technology in producing dental prostheses have been reached [52, 53]. However, Ayyildiz et al. concluded that the hardness, elastic modulus and strength of cobalt chromium alloys produced by laser AM is suitable for dental applications [54].

Mechanical Properties

Relative Density

Cobalt chromium alloys achieved a highest relative density of 99.94% by SLM [55].

Strength and Hardness

Table 2.3 shows the reported ultimate tensile strength for cobalt chromium, along with their respective yield strength and elongation for SLM and casting produced parts.

2.3.3 *Titanium-6Aluminium-4Vanadium*

Due to their biocompatibility, most research on Ti6Al4V is driven by its potential applications as body prostheses and implants. Ti6Al4V is of high interest because of its applications in biomedical industry [61, 62]. Ti6Al4V has been widely used for various orthopaedic implants due to its good biocompatibility, superior corrosion resistance and high mechanical strength [63, 64]. Moreover, Ti6Al4V has high specific strength and elastic moduli closer to bone than cobalt chromium alloys and stainless steel [65]. Performance requirements for implants made in Ti6Al4V are specified by ASTM standards, ISO standards and US FDA. Ti6Al4V components can be produced with various microstructures depending on the method used to process the alloy. For example, casting, wrought ingots and powder metallurgy give three different microstructures for Ti6Al4V. This is because for pure titanium, the microstructure is completely α. When pure titanium is alloyed with α and β stabilizers, β phase forms along the grain boundary. The percentage of α and β phases varies depending upon the processing conditions such as the temperature, cooling rates and degree of mechanical working.

Research into Ti6Al4V orthopaedic implants using SLM have been done by several groups. Lin et al. studied the mechanical properties of a Ti6Al4V cellular interbody fusion cage produced by SLM [17], Murr et al. investigated the microstructure and mechanical properties of SLM Ti6Al4V for biomedical applications [66]. Warnke et al. conducted cell experiments using SLM Ti6Al4V porous scaffolds. The scaffolds allow total overgrowth of osteoblasts (bone cells) [67]. Vandenbroucke & Kruth examined the dimensional accuracy of the SLM process for fabrication of dental frameworks using Ti6Al4V [13]. In recent years, various research groups have also carried out numerous studies on SLM of Ti6Al4V for body implants [19, 62, 63, 68–79]. In one of the latest studies, Biemond et al. examined the bone in-growth potential of trabecular-like implant surfaces produced by SLM of Ti6Al4V in goats. It is concluded that the SLM produced parts showed good bone in-growth characteristics after 15 weeks [80].

Mechanical Properties

Relative Density

SLM of Ti6Al4V has achieved positive results in relative density attained by various researchers. The highest relative density reported for SLM Ti6Al4V is 99.80% without any post-processing [13].

Strength and Hardness

Table 2.4 shows the reported ultimate tensile strength for Ti6Al4V, along with their respective yield strength and elongation for SLM and casting produced parts.

2.3.4 New Materials

Magnesium and its alloys have great potential for orthopaedic applications as they have been shown to have mechanical properties aligned to bone and induce no inflammatory response while being fully bio-resorbable. In addition, they are also osteoconductive, encourage bone growth and have a role in cell attachment [9, 86]. Ng

Table 2.4 Tensile strengths and micro-hardness of SLM and casted Ti6Al4V

Properties	SLM	Casting
Ultimate tensile strength	1250–1267 MPa [13, 81]	934–1173 MPa [82, 83]
Yield strength	1110–1125 MPa [13, 81]	862–999 MPa [82, 83]
Elongation	6–7% [13, 81]	6–7% [82, 83]
Micro-hardness	479–613 HV [23, 68]	294–360 HV [84, 85]

et al. examined the SLM of magnesium for applications in light-weight biodegradable and bio-resorbable magnesium parts for orthopaedic implants [87, 88]. The immediate challenge in processing magnesium using SLM is the severe oxidation and flammability of magnesium. A shielding box with gas inlets and outlets was designed to circulate inert argon gas inside the box within the build chamber. The argon gas minimised the risk of flame and oxidation during the process. Recently, research has also extended to include magnesium based alloys processed by SLM [89, 90].

There are recent studies that aim to replace Ti6Al4V with other titanium based alloys due to the toxicity of elements in Ti6Al4V [91, 92]. Chlebus et al. studied the possibility of using Ti6Al7Nb for medical implants as it replaces vanadium with niobium in its chemical composition. This alloy is found to have higher corrosion resistance and bio-tolerance compared to Ti6Al4V [93]. Further studies on Ti6Al7Nb titanium alloy were also done by Marcu et al. as endosseous implant [94]. Szymczyk et al. also examined cultured cell growth of staphylocuccus aureus on Ti6Al7Nb scaffolds which demonstrated the potential of this titanium alloy in biomedical applications [95]. SLM of Ti24Nb4Zr8Sn has been examined by Zhang et al. as an improvement over Ti6Al4V as it is an alloy with lower modulus of 53 $\pm$ 1 GPa. This results in a closer match of moduli between implant and surrounding bone. Due to the lower modulus, Ti24Nb4Zr8Sn has higher possibility of preventing bone resorption from stress shielding, which causes implant loosening [96].

Tantalum is a biomaterial with superior mechanical properties suitable for biomedical applications. With designed porosity, it can achieve an elastic modulus similar to that of bone, which minimises stress shielding [86]. Thijs et al. experimented using pure tantalum for fabrication using SLM and achieved a relative density of 99.6%. The yield strength of SLM tantalum is also reported to be higher than those obtained via conventional methods [97].

The capability of SLM to process powder mixtures has opened new and exciting material research opportunities. Several works have been reported on the production of several types of new powder mixture processed by SLM [98–102]. Vrancken et al. managed to create a novel metallic composite comprising of β titanium matrix and unmelted molybdenum particles using SLM [102]. In their work, a mixture of Ti6Al4V pre-alloyed powder and molybdenum powder were used. The resulting material has a Young's modulus of 73 $\pm$ 1 GPa, yield strength of 858 $\pm$ 16 MPa, ultimate tensile strength of 919 $\pm$ 10 MPa and elongation of 20.1 $\pm$ 2.0%.

2.4 Design Consideration for Metallic Implants and Scaffolds Using Selective Laser Melting

Metallic scaffolds offer advantages for two reasons. Firstly, the scaffolds allow immediate partial weight bearing with support of osteo-synthesis and long-term stability accompanied by its biocompatibility and bone in-growth ability into the open pores. Secondly, the risk of late fractures due to scaffold instability is almost negligible as long as the bone-scaffold interface is well incorporated with new bone formation [103]. However, metallic scaffolds are mostly not biodegradable and therefore, cannot be replaced by newly formed bone. The open porous structures must possess sufficient porosity for bone in-growth and nutrient supply while maintaining their load bearing capacity. Furthermore, there are still other considerations involved in manufacturing of these scaffolds such as the manufacturability of the designs and dimensional accuracy.

SLM can theoretically fabricate any shape of metal parts, the manufacturing quality can differ as the design and processing parameters change [104]. However, SLM offers the possibility to fully control both the geometrical and mechanical properties of scaffolds, which is important as both morphological and mechanical properties have an influence on in vivo and in vitro performance [105]. There is a need to balance between the manufacturability, mechanical properties and biological responses of these scaffolds produced by SLM.

2.4.1 Manufacturability of Metallic Scaffolds Using Selective Laser Melting

The manufacturability of metallic scaffolds using SLM depends on the design characteristics of the scaffolds, such as strut dimensions and unit cell shape. There is also a need to consider appropriate unit cell size, as overhanging struts in the cells can lead to deformation. Even though sacrificial support structures can be added to support the overhanging structures, thus preventing deformation, they are difficult to be removed from inside the complex cellular lattice structures [106]. This adds considerable constraints on manufacturing versatility. Furthermore, the manufacturability is also dependent on the SLM parameters such as laser spot size, laser power, laser scanning speed, hatch spacing and layer thickness [104]. The powder particle size used has effect on the manufacturability of the structures as well.

Zhang et al. studied the effect of hatch spacing on pore characteristics of Ti6Al4V structures fabricated using SLM. Pores are formed by varying the hatch spacing of the laser scans, instead of specific designs using CAD. The spot size of the laser used is 200 μm, hence, it is found that a hatch spacing of distance greater than the spot size is necessary for pores formation. Partially melted powder particles are also observed to adhere on the strut surfaces. It is advised that the pore diameters should be at least three times greater than the biggest powder particles for forming interconnected pores due to the accumulation of the powder [107]. In the same study, it is also concluded that the powder particle size has a significant influence on porosity formation and laser spot size directly determines the strut width. Qiu et al. investigated the influence of scanning speed and laser power on strut size, morphology and surface structures. It is found that higher laser powers lead to formation of thicker struts with larger deviation from the designed strut diameters. Higher laser power also leads to increased powder adhesion on the struts. However, the scanning speed only affect the strut diameter at lower end of the scanning speeds [108]. All these results coincide with study by Tsopanos et al. which state that the strut diameter is dependent on the energy applied on the powder, which is related to the laser power [109].

The designed and SLM produced lattice structures usually have disparity due to the following reasons: (1) an inadequately chosen beam offset that does not compensate for the laser spot size used, and hence, the melt pool formed during SLM differs from the desired cross section, (2) staircase effect, due to layer-by-layer fabrication, causes difference in designed and produced struts and (3) loose powder particles are likely to stick to the surface of parts, which leads to waviness and dimensional inaccuracy [110]. The causes of errors in dimensions are shown in Fig. 2.6.

Wang et al. concluded that powder adhesion is an inevitable problem in SLM process, especially in overhanging structures, which affect the manufacturability of metallic scaffolds by SLM, however, they can be controlled by optimizing design and process control [104]. Yan et al. also attributed powder adhesion due to balling phenomenon which give rise to beads being formed mainly on laser melted surfaces perpendicular to the building direction [106]. However, Abele et al. concluded that building orientation has no significant effect on the manufacturability of lattice structures by SLM [111], which implies that these powder adhesions have no significant effect on the short term mechanical properties. However, the powder adhesion can act as stress concentrators which facilitate fatigue crack initiations, affecting fatigue strength of the porous structures [112, 113]. Furthermore, since they are loosely bonded to the struts, they can be easily released into the biological system, causing inflammation [62]. A study of cobalt chrome molybdenum based super alloy, by Hazlehurst et al. also concluded that structural variation and heterogeneities can have detrimental effect on the stiffness of lattice structures manufactured using SLM [114]. Jet blasting or post-SLM sintering of the structures can lead to localized removal of these powder adhesion, with no effect on the macro-properties of the overall pore or strut network [115].

To fabricate lattice structures with precise dimensions, it is important to select appropriate processing conditions or to account for the oversizing of the struts compared to designed diameters.

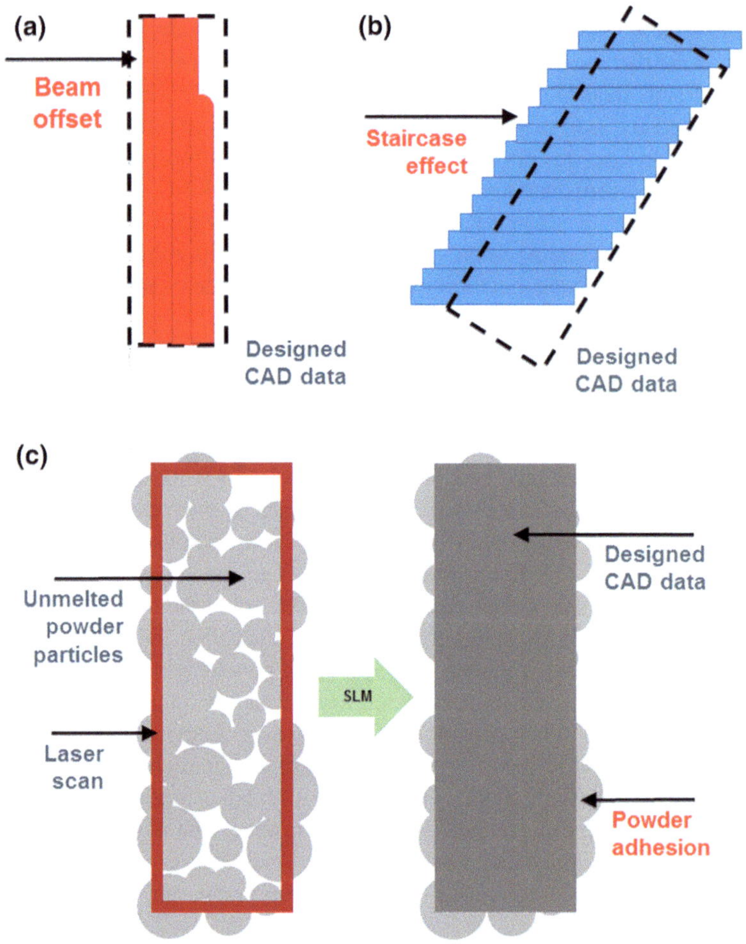

Fig. 2.6 Causes of errors between designed and fabricated lattice structures **a** beam offset **b** staircase effect **c** powder adhesion

2.4.2 Biological Compatibility of Metallic Scaffolds Fabricated Using Selective Laser Melting

Bone has a composite structure, containing about 45–60% minerals, 20–30% matrix and 10–20% water [86]. The human skeleton consists of cortical and trabecular bones. These two types of bone differ in terms of proportions of organic and inorganic materials, degree of porosity and organization [86]. Due to these differences, metallic implants and scaffolds require precise designs in order to mimic the bone properties closely [116].

Porosity is defined as the percentage of void space in a solid. It is a morphological property independent of the material [18]. A porous implant material with adequate pore structure and appropriate mechanical properties has long been sought as the ideal bone substitute as interconnected pores allow tissue in-growth and thus anchor the implant to the surrounding bone. The anchoring prevents loosening of the implant from the surround tissues [117]. Each porous material can have a combination of one to three types of pores. The types of pores include closed, through or blind pores. The closed pores are not accessible to fluids as they do not have opening at the surfaces of the materials. The blind pores open at surfaces of the material and terminate inside the material. The through pores are those that make possible the complete passageway of fluids from one surface to another of the material. The open porosity is made up of only through and blind pores [18]. Porosity that includes closed pores has a great impedance on mechanical properties of a material. Open porosity has its direct impact in the possibility of penetration desired and undesired fluids, cells and bacteria. Recent publications have shown that micro-porosity is an essential element in osteo-induction of biomaterials. However, there is a limit to which osteo-inductive potential can be increased by increasing the micro-porosity as micro-porosity affects the mechanical stability of the material surface. A mechanically stable surface of the material is needed in order to facilitate new bone formation [17, 18, 118, 119]. The pores of the structure also have to be interconnected in order to ensure bone in-growth [19, 119]. Porous structures are typically used as a coating over an underlying solid to provide surface for bone in-growth or as a structure to modify the modulus of the device allowing matching of the implant modulus to that of neighbouring bones [16].

The porosity of the implants and the pore architecture are critical in encouraging cell migration [120]. In addition to having suitable open pore size and porosity for cell attachment and proliferation, the implant must have sufficient mechanical strength to support physiological loadings *in vivo* [116, 121]. Another important consideration is the pores continuity as pores must be interconnected for cell migration [122]. It is important to note that although SLM can theoretically fabricate any porous structures, only structures with open pores can permit the removal of unmelted powder. A sample of metallic scaffold fabricated by SLM, with magnified struts under optical microscopy (OM), is shown in Fig. 2.7.

The biological performance of porous structures such as cell attachment, growth and differentiation are dependent on the pore shape, pore size and porosity [123]. To study such biological responses, many studies have been conducted using metallic scaffolds. Pore size and geometry play important roles in biological response both in bones and soft tissues. SLM allows production of predetermined pore sizes like those observed in bones. In addition, these porous structures can exert appropriate osteo-conductive properties and promote bone formation [77].

It is concluded that open porosity is essential for osteo-induction, osteo-conduction and osteo-integration. Fukuda et al. evaluated the effects of the inter-connected pore size on osteo-inductivity. For the range of 500–1200 μm, the best osteo-induction occurs for 500 μm pore size [74]. Porosity level of between 60 and 80% is preferred as at porosity above this level, the material is unlikely to possess sufficient mechanical properties for the intended application [10]. Barbas et al. stated

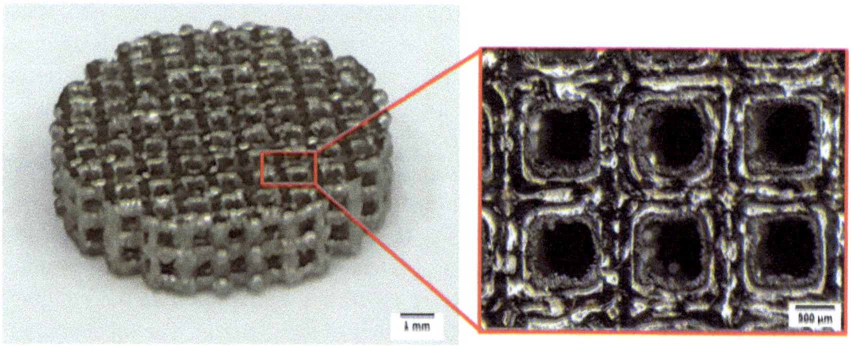

Fig. 2.7 Sample of metallic scaffold fabricated using SLM

that the size of the pores should range from 500 to 1500 μm and should not be less than 100 μm, as it would be impossible for bone colonization [19]. Mullen et al. further added that the optimal pore size for in-growth bone lies between 100 and 700 μm [10]. Pore size of less than 100 μm failed to support the capillary growth and did not allow the bone cells to bridge pores. However, pore size larger than 700 μm was able to promote bone in-growth but at reduced rate and volume [124]. Warnke et al. found that cells can occlude pores over two months if the pore size is 400–700 μm, but zero occlusion if the pore size is beyond 900 μm [67]. Pore overgrowth by osteoblasts can be desirable for bone formation within the pores and osteo-integration of the scaffold. Fukuda et al. obtained a similar result that osteo-induction is significant if the pore size is 500–600 μm [74]. In addition to pore size, pore shape may also affect cell proliferation and differentiation [125]. Recently, Wielding et al. studied the biological response of SLM produced scaffolds in sheep. The study showed an increase in bone mineral density between 12 and 24 weeks from 50 to 80% and concluded the capability of open porous titanium scaffolds for sufficient stabilization of large segmental bone defects [103].

The implication from these studies shows that SLM should fabricate porous structures that match at least three requirements: (1) the modulus of the target bone, (2) open pore size of approximately 500 μm and (3) compatible unit cell (pore) shape.

In addition to the designed pore characteristics, osteoblastic cell adhesion, growth and differentiation are related to surface energy and roughness [126]. Osteo-integration process is affected by surface conditions in terms of chemical and physical properties. Two conventional methods used to improve osteo-integration process are by applying coatings on the surface of the implants and by chemical treatment of the implant surface. A study by Fukuda et al. shows that chemically and heat treated implants exhibited better osteo-inductivity and induced ectopic bone growth [74]. Plasma spraying of hydroxyapatite is the most commonly used coating technique. Chemical treatments are carried out in order to obtain hydroxide (OH-) groups. By leveraging on SLM capabilities, surfaces with different roughness can be generated on the implants during the fabrication itself such that post-processing can be avoided.

Table 2.5 Biological responses of metallic scaffolds fabricated by SLM

Material	Pore size (μm)	Key findings	References
Ti6Al4V	240–730	Scaffolds provided enough mechanical support and encouraged bone formation with porosity of 68–88%	[77]
Ti6Al4V	450–1200	Porous structures allow total overgrowth of all pores of 500 μm and a significant proportion in the range of 500–600 μm by osteoblasts	[67]
Ti6Al4V	500 and 1000	Soaked in hydrofluoric acid for 10 min, rinsed with demineralized water and ethanol. Smaller pore size results in higher cell attachments due to lower permeability	[125]
Ti6Al4V	280–420	Surface coated with octacalcium phosphate or hydroxyapatite, both doped with Mg. Both as-build and surface modified implants are biocompatible and integrate with the bone with good bone-implant bonding. Bone volume was highest for implants coated with hydroxyapatite	[129]

By changing the process parameter settings, the surface roughness can be varied. The as-fabricated SLM part can have a rough surface with a Ra value of several microns, which is deemed to be more associated with bone-implant contact area and interface strength.

Pattanayak et al. reported that bone affinity of chemical and heat treated porous bodies with smoother surface were significantly higher than that of untreated scaffolds [63]. Rotaru et al. observed powder debris and little bone contact during a three month implantation [127]. In a study, porous titanium structures were fabricated using SLM and the surfaces were left untreated, sandblasted, or sandblasted/acid etched. It was observed that bone bridging was significantly increased in sand blasted acid etched scaffolds in the rabbit model. This indicates the needs for specially designed surface characteristic to improve osteo-conduction of the implants [128]. Surface treatment can decrease the ability of materials to form biofilm on the implants, which is very important to the production of medical implants [95]. Key findings on biological response of metallic scaffolds fabricated by SLM are tabulated in Table 2.5.

Fig. 2.8 Samples of compression test coupons fabricated by SLM

2.4.3 Mechanical Properties of Metallic Scaffolds Fabricated Using Selective Laser Melting

Likewise to the biological responses, the mechanical properties of cellular lattice structures are dependent on their morphological features such as the unit cell, pore size and porosity [123]. The mechanical properties such as surface quality is also dependent on the processing parameters and particle size distribution [46]. Cube or cylindrical samples are usually fabricated for mechanical testing based on ASTM E9 (Standard Test Methods of Compression Testing of Metallic Materials at Room Temperature) or ISO 13314:2011 (Mechanical testing of metals—Ductility testing—-Compression test for porous and cellular metals) [130]. Samples of test coupons of lattice structures fabricated using SLM are shown in Fig. 2.8.

Ahmadi et al. studied the analytical solutions and closed-form relations for predicting elastic modulus, Poison's ratio, critical buckling load and yield stress of cellular lattice structures [123]. It is concluded that finite element (FE) model made up of 14 repeating unit cells can be used to accurately predict the mechanical behaviours of cellular lattice structures of infinite repeating unit cells. Smith et al. also used FE model to predict the compressive response of lattice structures and found the results agreeable with the actual SLM produced structures despite using a different unit cell [131]. However, a study done by Bültmann et al. concluded that there is no scalability of mechanical properties on the struts produced by SLM [132]. Even though FE models are more accurate than mathematical model in predicting the mechanical properties, using FE models is more difficult as there is a need to create a new FE model for every new porous structures [69]. For small apparent density values, it is concluded that mathematical models are in good agreement with experimental results, however, for large apparent density values, the results from the mathematical model deviate significantly from actual experiments [123]. Ushijima et al. concluded that mathematical models are close to FE models and experimental results when the ratio of diameter to length is relatively small for the structures, however, FE models can be used for a wide range of diameter to length ratios [133]. To summarize, existing FE modelling technique can be used to accurately predict the mechanical properties of SLM produced structures, provided that the whole structure is simulated by the FE model.

Brenne et al. studied the compressive deformation behaviour of Ti6Al4V cellular lattice structures fabricated using SLM. It is found that heat treated samples, at 1050 °C for 2 h, under vacuum with subsequent furnace cooling, has significantly higher fatigue life under cyclic bending load [134]. Tsopanos et al. studied the compressive behaviour of stainless steel micro-lattice structures and concluded that low laser power results in low yield, ultimate tensile strength and elongation, which is attributed to significant number of unmelted powder [109]. Wauthle et al. studied the effects of build orientation on the mechanical properties of Ti6Al4V lattice structures with diamond unit cells, and concluded that the build orientation has significant influence on the mechanical properties. Structures built diagonally are inferior to both the horizontal and vertically orientated samples which have near identical properties [105]. This implies that horizontal struts should be avoided unless they can be supported with other struts. The results coincide with findings by Abele et al. [111]. Amin Yavari et al. studied the fatigue behaviour of Ti6Al4V porous lattice fabricated by SLM. It is found that the static mechanical properties of the porous structures are within reported range of mechanical properties of bone, however, the normalized endurance limits of the tested structures are lower than some other porous structures manufactured using other techniques [135]. Key findings of mechanical properties of metallic lattice structures fabricated using SLM are tabulated in Tables 2.6 and 2.7.

2.5 Summary

With the advancement in AM technologies, it is now able to fabricate fully functional metallic parts directly. SLM provides the opportunities to mass customize implants due to their ability to fabricate parts with complex and intrinsic designs that can be specific to individual patients.

To summarize, all the above studies show the immense potential of SLM to become the preferred method for producing medical implants in future. However, there are still research challenges that need to be overcome, in areas such as:

(1) Material Research

Current metallic biomaterials such as 316L stainless steel, cobalt chromium and Ti6Al4V have been widely adopted for use in orthopaedic industry. However, these materials have moduli that are much higher than that of human bones (1.0–25.0 GPa). Mismatch of modulus with the neighbouring bones will result in stress shielding. To

Table 2.6 Compression properties of Ti6Al4V scaffolds fabricated by SLM

Test description	Key findings	References
Uniaxial loading was carried out using a servo-hydraulic test rig with a maximal load capacity of ±15 kN	Steep drop in stresses occurs at strain of about 5% due to failure of struts along the entire plane of the cube. The drop is steeper for the as-built samples due to inferior ductility. Higher maximum stresses is achieved by the heat treated sample	[134]
Tested in accordance to ISO 13314:2011	As the number of unit cells used in x-, y-, and z-directions increased from 5 to 20, the cellular structure exhibited a stiffer response. Comparison between the mathematical, numerical, and experimental results shows that for small values of the apparent density, all methods yield very similar results. As the apparent density increases, Young's moduli estimated using the FE model match the experimental results very well even for large apparent density values	[123]
Mechanical testing with a 100 kN load cell at a strain rate of 10^{-3} s^{-1}	Different hatch spacing results in yield strength between 467 and 862 MPa, and Young's modulus in the range of 16–85 GPa. There is no significant influence of hatch spacing on failure mechanism, which is due to adiabatic shear band	[107]

minimise such adverse effect, there is a need to develop new materials that have the strength to act as replacement of bones for load bearing supports and yet has modulus as close as possible to that of bones.

(2) Designs of Orthopaedic Implants

There is now more design freedom for implants due to the growing prominence of SLM in this field. Despite these advantages, SLM still have limitations such as the limited pool of materials that can be processed by them as well as the lack of understanding in SLM constraints for lattice fabrication. Furthermore, the dimensional accuracy of the designed cellular structures and the entrapment of powder within them for orthopaedic implants proved to be challenging. However, these limitations can be overcome soon by further improvement in SLM. Improvement to the process can be achieved by more in-depth study of the process such as the interaction of processing parameter with powders to control the fabrication of cellular lattice structures.

Table 2.7 Tensile, bending and fatigue properties of Ti6Al4V scaffolds fabricated by SLM

Mechanical property	Test description	Key findings	References
Tensile	Uniaxial loading was carried out using a MTS servo-hydraulic test rig with a maximal load capacity of ±15 kN	Annealed samples are able to bear higher load at early stage of deformation. Heat treated samples have high ductility, allowing the struts to align along the loading axis which improves their load carrying capacity	[134]
Fatigue	Cyclic test with force control was used, applying a peak load of 25% of the maximum load reached by the as-built specimens under monotonic tensile and bending load respectively	Heat treated sample have higher fatigue life. The shortening of the sample (in compressive force) was related to strain accumulation and a reduction of stiffness due to crack initiation and growth within the struts	[134]
Bending	Four-point-bending test with rolls of 16 mm diameter were installed. Distance between the two upper rolls is 35 mm while the distance between the two lower rolls was 70 mm. This setup was mounted to a testing system capable of handling ±15 kN	Annealed samples have higher ductility and more intact struts. This results in higher stiffness	[134]

References

1. F. Watari, A. Yokoyama, M. Omori, T. Hirai, H. Kondo, M. Uo, T. Kawasaki, Biocompatibility of materials and development to functionally graded implant for bio-medical application. Compos. Sci. Technol. **64**, 893–908 (2004)
2. A. Sadollah, A. Bahreininejad, Optimum gradient material for a functionally graded dental implant using metaheuristic algorithms. J. Mech. Behav. Biomed. Mater. **4**, 1384–1395 (2011)
3. G. Manivasagam, D. Dhinasekaran, A. Rajamanickam, Biomedical implants: corrosion and its prevention—a review. Recent Pat. Corros. Sci. **2**, 40–54 (2010)
4. H. Rotaru, R. Schumacher, S.-G. Kim, C. Dinu, Selective laser melted titanium implants: a new technique for the reconstruction of extensive zygomatic complex defects, Maxillofacial Plastic and Reconstructive Surgery, 37 (2015)
5. S. Merkt, A. Kleyer, A.J. Hueber, The additive manufacture of patient-tailored finger implants. Laser Technik Journal **11**, 54–56 (2014)
6. T. Kokubo, M. Hanakawa, M. Kawashita, M. Minoda, T. Beppu, T. Miyamoto, T. Nakamura, Apatite formation on non-woven fabric of carboxymethylated chitin in SBF. Biomaterials **25**, 4485–4488 (2004)
7. K.S. Katti, Biomaterials in total joint replacement, Colloids and surfaces. B, Biointerfaces **39**, 133–142 (2004)

8. T. Kokubo, H.-M. Kim, F. Miyaji, H. Takadama, T. Miyazaki, Ceramic-metal and ceramic-polymer composites prepared by biomimetic process. Compos. Part A: Appl. Sci. Manuf. **30**, 405–409 (1999)
9. M.P. Staiger, A.M. Pietak, J. Huadmai, G. Dias, Magnesium and its alloys as orthopedic biomaterials: a review. Biomaterials **27**, 1728–1734 (2006)
10. L. Mullen, R.C. Stamp, P. Fox, E. Jones, C. Ngo, C.J. Sutcliffe, Selective laser melting: a unit cell approach for the manufacture of porous, titanium, bone in-growth constructs, suitable for orthopedic applications. II. Randomized structures. J. Biomed. Mater. Res. Part B: Appl. Biomater. **92**, 178–188 (2010)
11. T. Traini, C. Mangano, R.L. Sammons, F. Mangano, A. Macchi, A. Piattelli, Direct laser metal sintering as a new approach to fabrication of an isoelastic functionally graded material for manufacture of porous titanium dental implants. Dental Mater. **24**, 1525–1533 (2008)
12. E. Marin, L. Fedrizzi, L. Zagra, Porous metallic structures for orthopaedic applications: a short review of materials and technologies. Eur. Orthop. Traumatol. **1**, 103–109 (2010)
13. B. Vandenbroucke, J.-P. Kruth, Selective laser melting of biocompatible metals for rapid manufacturing of medical parts. Rapid Prototyp. J. **13**, 196–203 (2007)
14. G. Lewis, Properties of open-cell porous metals aand alloys for orthopaedic applications. J. Mater. Sci.: Mater. Med. **24**, 2293–2325 (2013)
15. B. Gorny, T. Niendorf, J. Lackmann, M. Thoene, T. Troester, H.J. Maier, In situ characterization of the deformation and failure behavior of non-stochastic porous structures processed by selective laser melting. Mat. Sci. Eng.: A **528**, 7962–7967 (2011)
16. R. Stamp, P. Fox, W. O'Neill, E. Jones, C. Sutcliffe, The development of a scanning strategy for the manufacture of porous biomaterials by selective laser melting. J. Mater. Sci.: Mater. Med. **20**, 1839–1848 (2009)
17. C.Y. Lin, T. Wirtz, F. LaMarca, S.J. Hollister, Structural and mechanical evaluations of a topology optimized titanium interbody fusion cage fabricated by selective laser melting process. J. Biomed. Mater. Res. Part A **83**, 272–279 (2007)
18. M. Mour, D. Das, T. Winkler, E. Hoenig, G. Mielke, M.M. Morlock, A.F. Schilling, Adv. Porous Biomater. Dental Orthop. Appl. Materials **3**, 2947–2974 (2010)
19. A. Barbas, A.-S. Bonnet, P. Lipinski, R. Pesci, G. Dubois, Development and mechanical characterization of porous titanium bone substitutes. J. Mech. Behav. Biomed. Mater. **9**, 34–44 (2012)
20. X. Fan, B. Feng, J. Weng, J. Wang, X. Lu, Processing and properties of porous titanium with high porosity coated by bioactive titania nanotubes. Mater. Lett. **65**, 2899–2901 (2011)
21. H.K. Rafi, N.V. Karthik, H. Gong, T.L. Starr, B.E. Stucker, Microstructures and mechanical properties of Ti6Al4V parts fabricated by selective laser melting and electron beam melting. J. Mater. Eng. Perform. **22**, 3872–3883 (2013)
22. C.K. Chua, K.F. Leong, *3D Printing and Additive Manufacturing: Principles and Applications*, 5th edn. (World Scientific Publishing Co. Pte. Ltd, Singapore, 2017)
23. L. Thijs, F. Verhaeghe, T. Craeghs, J. Van Humbeeck, J.P. Kruth, A study of the microstructural evolution during selective laser melting of Ti-6Al-4V. Acta Mater. **58**, 3303–3312 (2010)
24. T.G. Spears, S.A. Gold, In-process sensing in selective laser melting (SLM) additive manufacturing. Integr. Mater. Manuf. Innov. **5** (2016)
25. I. Yadroitsev, P. Bertrand, I. Smurov, Parametric analysis of selective laser melting process. Appl. Surf. Sci. **253**, 8064–8069 (2007)
26. Z.H. Liu, D.Q. Zhang, S.L. Sing, C.K. Chua, L.E. Loh, Interfacial characterization of SLM parts in multi-material processing: metallurgical diffusion between 316L stainless steel and C18400 copper alloy. Mater. Charact. **94**, 116–125 (2014)
27. K. Kunze, T. Etter, J. Grässlin, V. Shklover, Texture, anisotropy in microstructure and mechanical properties of IN718LC alloy processed by selective laser melting (SLM). Mater. Sci. Eng.: A **620**, 213–222 (2015)
28. Wei, Q., Shuai, L., Han, C., Li, W., Cheng, L., Hao, L., Shi, Y.: Selective laser melting of stainless-steel/nano-hydroxyapatite composites for medical applications: Microstructure, element distribution, crack and mechanical properties. J. Mater. Process. Technol. (2015)

29. K.G. Prashanth, H. Shakur Shahabi, H. Attar, V.C. Srivstava, N. Ellendt, V. Uhlenwinkel, J. Eckert, S. Scudino, Production of high strength Al85Nd8Ni5Co2 alloy by selective laser melting. Addit. Manuf. **6**, 1–5 (2015)
30. D. Hu, Y. Wang, D. Zhang, L. Hao, J. Jiang, Z. Li, Y. Chen, Experimental investigation on selective laser melting of bulk net-shape pure magnesium. Mater. Manuf. Process (2015)
31. W.Y. Yeong, C.Y. Yap, M. Mapar, C.K. Chua, State-of-the-art review on selective laser melting of ceramics. High Value Manuf.: Adv. Res. Virtual Rapid Prototyp. 65–70 (2013)
32. D. Dai, D. Gu, Tailoring surface quality through mass and momentum transfer modeling using a volume of fluid method in selective laser melting of TiC/AlSi10Mg powder. Int. J. Mach. Tools Manuf. **88**, 95–107 (2015)
33. P. Yuan, D. Gu, Molten pool behaviour and its physical mechanism during selective laser melting of TiC/AlSi10Mg nanocomposites: simulation and experiments. J. Phys. D: Appl. Phys. **48** (2015)
34. H. Attar, K.G. Prashanth, A.K. Chaubey, M. Calin, L.C. Zhang, S. Scudino, J. Eckert, Comparison of wear properties of commercially pure titanium prepared by selective laser melting and casting processes. Materl. Lett. **142**, 38–41 (2015)
35. G. Kasperovich, J. Hausmann, Improvement of fatigue resistance and ductility of Ti6Al4V processed by selective laser melting. J. Mater. Process. Technol. **220**, 202–214 (2015)
36. S.L. Sing, L.P. Lam, D.Q. Zhang, Z.H. Liu, C.K. Chua, Interfacial characterization of SLM parts in multi-material processing: intermetallic phase formation between AlSi10Mg and C18400 copper alloy. Mater. Charact. **107**, 220–227 (2015)
37. Y.Q. Yang, J.B. Lu, Z.Y. Luo, D. Wang, Accuracy and density optimization in directly fabricating customized orthodontic production by selective laser melting. Rapid Prototyp. J. **18**, 482–489 (2012)
38. R.D. Li, J.H. Liu, Y.S. Shi, M.Z. Du, Z. Xie, 316L stainless steel with gradient porosity fabricated by selective laser melting. J. Mater. Eng. Perform. **19**, 666–671 (2010)
39. R. Bibb, D. Eggbeer, R. Williams, Rapid manufacture of removable partial denture frameworks. Rapid Prototyp. J. **12**, 95–99 (2006)
40. J.-P. Kruth, P. Mercelis, J. Van Varenbergh, L. Froyen, M. Rombouts, Binding mechanisms in selective laser sintering and selective laser melting. Rapid Prototyp. J. **11**, 26–36 (2005)
41. R. Bibb, D. Eggbeer, P. Evans, A. Bocca, A. Sugar, Rapid manufacture of cutom-fitting surgical guides. Rapid Prototyp. J. **15**, 346–354 (2009)
42. M. Wehmoller, P.H. Warnke, C. Zilian, H. Eufinger, Implant design and production - a new approach by selective laser melting, in *CARS 2005: Computer Assisted Radiology and Surgery*, ed. by H.U. Lemke, K. Inamura, K. Doi, M.W. Vannier, A.G. Farman (Elsevier Science Bv, Amsterdam, 2005), pp. 690–695
43. I. Tolosa, F. Garciandía, F. Zubiri, F. Zapirain, A. Esnaola, Study of mechanical properties of AISI 316 stainless steel processed by "selective laser melting", following different manufacturing strategies. Int. J. Adv. Manuf. Technol. **51**, 639–647 (2010)
44. E. Yasa, J. Deckers, J.-P. Kruth, The investigation of the influence of laser re-melting on density, surface quality and microstucture of selective laser melting parts. Rapid Prototyp. J. **17**, 312–327 (2011)
45. J. Delgado, J. Ciurana, L. Serenó, Comparison of forming manufacturing processes and selective laser melting technology based on the mechanical properties of products. Virtual Phys. Prototyp. **6**, 167–178 (2011)
46. A.B. Spierings, N. Herres, G. Levy, Influence of the particle size distribution on surface quality and mechanical properties in AM steel parts. Rapid Prototyp. J. **17**, 195–202 (2011)
47. L. Zhang, J. Wang, Effect of temperature and loading mode on environmentally assisted crack growth of a forged 316L SS in oxygenated high-temperature water. Corros. Sci. **87**, 278–287 (2014)
48. S. Venugopal, M. Vasudevan, S. Venugopal, P.V. Sivaprasad, S.K. Jha, P. Pandey, S.L. Mannan, Y.V.R.K. Prasad, Industrial validation of processing maps of 316L stainless steel using hot forging, rolling and extrusion. Mater. Sci. Technol. **12**, 955–962 (1996)

49. A.B. Spierings, T.L. Starr, K. Wegener, Fatigue performance of additive manufactured metallic parts. Rapid Prototyp. J. **19**, 88–94 (2013)
50. T.F. Kong, L.C. Chan, T.C. Lee, Experimental study of effects of process parameters in forge-welding bimetallic materials: AISI 316L stainless steel and 6063 aluminium alloy. Strain **45**, 373–379 (2009)
51. J. Delgado, J. Ciurana, C.A. Rodriguez, Influence of process parameters on part quality and mechanical properties for DMLS and SLM with iron-based materials. Int. J. Adv. Manuf. Technol. **60**, 601–610 (2012)
52. K.B. Kim, W.C. Kim, H.Y. Kim, J.H. Kim, An evaluation of marginal fit of three-unit fixed dental prostheses fabricated by direct metal laser sintering system. Dental Mater. **29**, 91–96 (2013)
53. R.C. Oyague, A. Sanchez-Turrion, J.F. Lopez-Lozano, J. Montero, A. Albaladejo, M.J. Suarez-Garcia, Evaluation of fit of cement-retained implant-supported 3-unit structures fabricated with direct metal laser sintering and vacuum casting techniques. Odontology **100**, 249–253 (2012)
54. S. Ayyildiz, E.H. Soylu, S. Ide, S. Kilic, C. Sipahi, B. Piskin, H.S. Gokce, Annealing of Co-Cr dental alloy: effects on nanostructure and Rockwell hardness. J. Adv. Prosthodont. **5**, 471–478 (2013)
55. C. Sanz, V.G. Navas, Structural integrity of direct metal laser sintered parts subjected to thermal and finishing treatments. J. Mater. Process. Technol. **213**, 2126–2136 (2013)
56. A. Takachi, T. Suyalatu, N. Nakamoto, N. Joko, Y. Nomura, S. Tsutsumi, H. Migita, S. Doi, A. Kurosu, N. Chiba, Y. Wakabayashi, T. Igarashi, Hanawa, Microstructures and mechanical properties of Co–29Cr–6Mo alloy fabricated by selective laser melting process for dental applications. J. Mech. Behav. Biomed. Mater. **21**, 67–76 (2013)
57. M. Averyanova, P. Bertrand, B. Verquin, Manufacture of Co-Cr dental crowns and bridges by selective laser melting technology. Virtual Phys. Prototyp. **6**, 173–185 (2011)
58. L. Wu, H. Zhu, X. Gai, Y. Wang, Evaluation of the mechanical properties and porcelain bond strength of cobalt-chromium dental alloy fabricated by selective laser melting. J. Prosthet. Dentist. **111**, 51–55 (2013)
59. X.Z. Xin, J. Chen, N. Xiang, B. Wei, Surface properties and corrosion behavior of Co-Cr alloy fabricated with selective laser melting technique. Cell Biochem. Biophys. **67**, 983–990 (2013)
60. B. Henriques, D. Soares, F.S. Silva, Microstructure, hardness, corrosion resistance and porcelain shear bond strength comparison between cast and hot pressed CoCrMo alloy for metal–ceramic dental restorations. J. Mech. Behav. Biomed. Mater. **12**, 83–92 (2012)
61. E. Marin, M. Pressacco, S. Fusi, A. Lanzutti, S. Turchet, L. Fedrizzi, Characterization of grade 2 commercially pure Trabecular Titanium structures. Mater. Sci. Eng.: C **33**, 2648–2656 (2013)
62. E. Sallica-Leva, A.L. Jardini, J.B. Fogagnolo, Microstructure and mechanical behavior of porous Ti-6Al-4V parts obtained by selective laser melting. J. Mech. Behav. Biomed. Mater. **26**, 98–108 (2013)
63. D.K. Pattanayak, A. Fukuda, T. Matsushita, M. Takemoto, S. Fujibayashi, K. Sasaki, N. Nishida, T. Nakamura, T. Kokubo, Bioactive Ti metal analogous to human cancellous bone: Fabrication by selective laser melting and chemical treatments. Acta Biomaterialia **7**, 1398–1406 (2011)
64. A. Fukuda, M. Takemoto, K. Tanaka, S. Fujibayashi, D.K. Pattanayak, T. Matsushita, K. Sasaki, N. Nishida, T. Kokubo, T. Nakamura, Bone ingrowth into pores of lotus Stem-Type bioactive titanium implants fabricated using rapid prototyping technique. Bioceram. Dev. Appl. **1**, 1–3 (2010)
65. F. Abe, E.C. Santos, Y. Kitamura, K. Osakada, M. Shiomi, Influence of forming conditions on the titanium model in rapid prototyping with the selective laser melting process. Proc. Inst. Mech. Eng. Part C—J. Mech. Eng. Sci. **217**, 119–126 (2003)
66. L.E. Murr, S.A. Quinones, S.M. Gaytan, M.I. Lopez, A. Rodela, E.Y. Martinez, D.H. Hernandez, E. Martinez, F. Medina, R.B. Wicker, Microstructure and mechanical behavior of

Ti-6Al-4V produced by rapid-layer manufacturing, for biomedical applications. J. Mech. Behavi. Biomed. Mater. **2**, 20–32 (2009)
67. P.H. Warnke, T. Douglas, P. Wollny, E. Sherry, M. Steiner, S. Galonska, S.T. Becker, I.N. Springer, J. Wiltfang, S. Sivananthan, Rapid prototyping: porous titanium alloy scaffolds produced by selective laser melting for bone tissue engineering. Tissue Eng.—Part C: Methods **15**, 115–124 (2009)
68. L.S. Bertol, W. Kindlein, F.P. da Silva, C. Aumund-Kopp, Medical design: direct metal laser sintering of Ti-6Al-4V. Mater. Des. **31**, 3982–3988 (2010)
69. G. Campoli, M.S. Borleffs, S. Amin Yavari, R. Wauthle, H. Weinans, A.A. Zadpoor, Mechanical properties of open-cell metallic biomaterials manufactured using additive manufacturing. Mater. Des. **49**, 957–965 (2013)
70. L. Ciocca, M. Fantini, F. De Crescenzio, G. Corinaldesi, R. Scott, Direct metal laser sintering (DMLS) of a customized titanium mesh for prosthetically guided bone regeneration of atrophic maxillary arches. Med. Biol. Eng. Comput. **49**, 1347–1352 (2011)
71. N. de Beer, A. van der Merwe, Patient-specific intervertebral disc implants using rapid manufacturing technology. Rapid Prototyp. J. **19**, 126–139 (2013)
72. C. Emmelmann, P. Scheinemann, M. Munsch, V. Seyda, Laser Additive Manufacturing of Modified Implant Surfaces with Osseointegrative Characteristics, in ed. by M.Z.A.M. Schmidt, T. Graf, A. Ostendorf, Lasers in Manufacturing 2011: Proceedings of the Sixth International WLT Conference on Lasers in Manufacturing (Elsevier Science Bv., Amsterdam, 2011), pp. 375–384
73. M. Figliuzzi, F. Mangano, C. Mangano, A novel root analogue dental implant using CT scan and CAD/CAM: selective laser melting techonology. Int. J. Oral Maxillofac. Surg. **41**, 858–862 (2012)
74. A. Fukuda, M. Takemoto, T. Saito, S. Fujibayashi, M. Neo, D.K. Pattanayak, T. Matsushita, K. Sasaki, N. Nishida, T. Kokubo, T. Nakamura, Osteoinduction of porous Ti implants with a channel structure fabricated by selective laser melting. Acta Biomaterialia **7**, 2327–2336 (2011)
75. W.S. Lin, T.L. Starr, B.T. Harris, A. Zandinejad, D. Morton, Additive manufacturing technology (Direct Metal Laser Sintering) as a novel approach to fabricate functionally graded titanium implants: preliminary investigation of fabrication parameters. Int. J. Oral Maxillofac. Implant. **28**, 1490–1495 (2013)
76. M. Salmi, J. Tuomi, K.S. Paloheimo, S. Bjorkstrand, M. Paloheimo, J. Salo, R. Kontio, K. Mesimaki, A.A. Makitie, Patient-specific reconstruction with 3D modeling and DWLS additive manufacturing. Rapid Prototyp. J. **18**, 209–214 (2012)
77. J. Van der Stok, O.P. Van der Jagt, S. Amin Yavari, M.F.P. De Haas, J.H. Waarsing, H. Jahr, E.M.M. Van Lieshout, P. Patka, J.A.N. Verhaar, A.A. Zadpoor, H. Weinans, Selective laser melting-produced porous titanium scaffolds regenerte bone in critical size cortical bone defects. J. Orthop. Res. **31**, 792–799 (2013)
78. D.M. Xiao, Y.Q. Yang, X.B. Su, D. Wang, J.F. Sun, An integrated approach of topology optimized design and selective laser melting process for titanium implants materials. Bio-Med. Mater. Eng. **23**, 433–445 (2013)
79. Z. Zhang, D. Jones, S. Yue, P.D. Lee, J.R. Jones, S.C. J., E. Jones, Hierarchical tailoring of strut architecture to control permeability of additive manufactured titanium implants. Mater. Sci. Eng. C: Mater. Biol. Appl. **33**, 4055–4062 (2013)
80. J.E. Biemond, G. Hannink, N. Verdonschot, P. Buma, Bone ingrowth potential of electron beam and selective laser melting produced trabecular-like implant surfaces with and without a biomimetic coating. J. Mater. Sci.: Mater. Med. **24**, 745–753 (2013)
81. B. Vrancken, L. Thijs, J.P. Kruth, J. Van Humbeeck, Heat treatment of Ti6Al4V produced by selective laser melting: microstructure and mechanical properties. J. Alloy. Compd. **541**, 177–185 (2012)
82. Z.J. Wei, L. Cao, H.W. Wang, C.M. Zou, Microstructure and mechanical properties of TiC/Ti–6Al–4V composites processed by in situ casting route. Mater. Sci. Technol. **27**, 1321–1327 (2011)

83. W.F. Ho, A comparison of tensile properties and corrosion behavior of cast Ti-7.5Mo with c.p. Ti, Ti-15Mo and Ti-6Al-4V alloys. J. Alloy. Compd. **464**, 580–583 (2008)
84. W.F. Ho, C.P. Ju, J.H. C. Lin, Structure and properties of cast binary Ti-Mo alloys. Biomaterials **20**, 2115–2122 (1999)
85. M.T. Jovanovic, S. Tadic, S. Zec, Z. Miskovic, I. Bobic, The effect of annealing temperatures and cooling rates on microstructure and mechanical properties of investment cast Ti-6Al-4V alloy. Mater. Des. **27**, 192–199 (2006)
86. K. Alvarez, H. Nakajima, Metallic scaffolds for bone regeneration. Materials **2**, 790–832 (2009)
87. C.C. Ng, M.M. Savalani, H.C. Man, I. Gibson, Layer manufacturing of magnesium and its alloy structures for future applications. Virtual Phys. Prototyp. **5**, 13–19 (2010)
88. C.C. Ng, M.M. Savalani, M.L. Lau, H.C. Man, Microstructure and mechanical properties of selective laser melted magnesium. Appl. Surf. Sci. **257**, 7447–7454 (2011)
89. C. Shuai, C. He, P. Feng, W. Guo, C. Gao, P. Wu, Y. Yang, S. Bin, Biodegradation mechanisms of selective laser-melted Mg–xAl–Zn alloy: grain size and intermetallic phase. Virtual Phys. Prototyp. **13**, 59–69 (2018)
90. T. Long, X. Zhang, Q. Huang, L. Liu, Y. Liu, J. Ren, Y. Yin, D. Wu, H. Wu, Novel Mg-based alloys by selective laser melting for biomedical applications: microstructure evolution, microhardness and in vitro degradation behaviour. Virtual Phys. Prototyp. **13**, 71–81 (2018)
91. M. Niinomi, Mechanical properties of biomedical titanium alloys. Mater. Sci. Eng.: A **243**, 231–236 (1998)
92. M. Niinomi, D. Kuroda, K.-I. Fukunaga, M. Morinaga, Y. Kato, T. Yashiro, A. Suzuki, Corrosion wear fracture of new β type biomedical titanium alloys. Mater. Sci. Eng.: A **263**, 193–199 (1999)
93. E. Chlebus, B. Kuznicka, T. Kurzynowski, B. Dybala, Microstructure and mechanical behaviour of Ti-6Al-7Nb alloy produced by selective laser melting. Mater. Charact. **62**, 488–495 (2011)
94. T. Marcu, M. Todea, L. Maines, D. Leordean, P. Berce, C. Popa, Metallurgical and mechanical characterisation of titanium based materials for endosseous applications obtained by selective laser melting. Powder Metall. **55**, 309–314 (2012)
95. P. Szymczyk, A. Junka, G. Ziolkowski, D. Smutnicka, M. Bartoszewicz, E. Chlebus, The ability of S.aureus to form biofilm on the Ti-6Al-7Nb scaffolds produced by Selective Laser Melting and subjected to the different typs of surface modifications. Acta Bioeng. Biomech. **15**, 69–76 (2013)
96. L.C. Zhang, D. Klemm, J. Eckert, Y.L. Hao, T.B. Sercombe, Manufacture by selective laser melting and mechanical behavior of a biomedical Ti-24Nb-4Zr-8Sn alloy. Scripta Materialia **65**, 21–24 (2011)
97. L. Thijs, M.L. Montero Sistiaga, R. Wauthle, Q. Xie, J.-P. Kruth, J. Van Humbeeck, Strong morphological and crystallographic texture and resulting yield strength anisotropy in selective laser melted tantalum. Acta Materialia **61**, 4657–4668 (2013)
98. H. Tan, J. Chen, F. Zhang, X. Lin, W. Huang, Microstructure and mechanical properties of laser solid formed Ti-6Al-4V from blended elemental powders. Rare Metal Mater. Eng. **38**, 574–578 (2009)
99. D. Gu, Y.-C. Hagedorn, W. Meiners, K. Wissenbach, R. Poprawe, Selective Laser Melting of in-situ TiC/Ti5Si3 composites with novel reinforcement architecture and elevated performance. Surf. Coat. Technol. **205**, 3285–3292 (2011)
100. D. Gu, W. Meiners, Microstructure characteristics and formation mechanisms of in situ WC cemented carbide based hardmetals prepared by selective laser melting. Mater. Sci. Eng.: A **527**, 7585–7592 (2010)
101. D. Gu, G. Meng, C. Li, W. Meiners, R. Poprawe, Selective laser melting of TiC/Ti bulk nanocomposites: Influence of nanoscale reinforcement. Scripta Materialia **67**, 185–188 (2012)
102. B. Vrancken, L. Thijs, J.-P. Kruth, J. Van Humbeeck, Microstructure and mechanical properties of a novel β titanium metallic composite by selective laser melting. Acta Materialia **68**, 150–158 (2014)

103. J. Wielding, T. Lindner, P. Bergschmidt, R. Bader, Biomechanical stability of novel mechanically adapted open-porous titanium scaffolds in metatarsal bone defects of sheep. Biomaterials **46**, 35–47 (2015)
104. D. Wang, Y. Yang, R. Liu, D. Xiao, J. Sun, Study on the designing rules and processability of porous structure based on selective laser melting (SLM). J. Mater. Process. Technol. **213**, 1734–1742 (2013)
105. R. Wauthle, B. Vrancken, B. Beynaerts, K. Jorissen, J. Schrooten, J.-P. Kruth, J. Van Humbeeck, Effects of build orientation and heat treatment on the microstructure and mechanical properties of selective laser melted Ti6Al4V lattice structures. Addit. Manuf. **5**, 77–84 (2015)
106. C. Yan, L. Hao, A. Hussein, D. Raymont, Evaluations of cellular lattice structures manufactured using selective laser melting. Int. J. Mach. Tools Manuf. **62**, 32–38 (2012)
107. S. Zhang, Q. Wei, L. Cheng, S. Li, Y. Shi, Effects of scan line spacing on pore characteristics and mechanical properties of porous Ti6Al4V implants fabricated by selective laser melting. Mater. Des. **63**, 185–193 (2014)
108. C. Qiu, S. Yue, N.J.E. Adkins, M. Ward, H. Hassanin, P.D. Lee, P.J. Withers, M.M. Attallah, Influence of processing condiions on strut structure and compressive properties of cellular lattice structures fabricated by selective laser melting. Mater. Sci. Eng. A **628**, 188–197 (2015)
109. S. Tsopanos, R.A.W. Mines, S. Mckown, Y. Shen, W.J. Cantwell, C.J. Sutcliffe, The influence of processing parameters on the mechanical properties of selectively laser melted stainless steel microlattice structures. J. Manuf. Sci. Eng. **132**, 041011-041011–041012 (2010)
110. S. Van Bael, G. Kerckhofs, M. Moesen, G. Pyka, J. Schrooten, J.P. Kruth, Micro-CT-based improvement of geometrical and mechanical controllability of selective laser melted Ti6Al4V porous structures. Mater. Sci. Eng. A **528**, 7423–7431 (2011)
111. E. Abele, H.A. Stoffregen, K. Klimkeit, H. Hoche, M. Oechsner, Optimisation of process parameters for lattice structures. Rapid Prototyp. J. **21**, 117–127 (2015)
112. S.J. Li, L.E. Murr, X.Y. Cheng, Z.B. Zhang, Y.L. Hao, R. Yang, F. Medina, R.B. Wicker, Compression fatigue behavior of Ti–6Al–4V mesh arrays fabricated by electron beam melting. Acta Materialia **60**, 793–802 (2012)
113. N.W. Hrabe, P. Heinl, B. Flinn, C. Korner, R.K. Bordia, Compression-compression fatigue of selective electron beam melted cellular titanium (Ti-6Al-4V). J. Biomed. Mater. Res.—Part B Appl. Biomater. **99B**, 313–320 (2011)
114. K. Hazlehurst, C.J. Wang, M. Stanford, Evaluation of the stiffness characteristics of square pore CoCrMo cellular structures manufactured using laser melting technology for potential orthopaedic applications. Mater. Des. **51**, 949–955 (2013)
115. T.B. Kim, S. Yue, Z. Zhang, E. Jones, J.R. Jones, P.D. Lee, Additive manufactured porous titanium structures: through-process quantification of pore and strut networks. J. Mater. Process. Technol. **214**, 2706–2715 (2014)
116. N. Sudarmadji, J.Y. Tan, K.F. Leong, C.K. Chua, Y.T. Loh, Investigation of the mechanical properties and porosity relationships in selective laser-sintered polyhedral for functionally graded scaffolds. Acta Biomaterialia **7**, 530–537 (2011)
117. M. Takemoto, S. Fujibayashi, M. Neo, J. Suzuki, T. Kokubo, T. Nakamura, Mechanical properties and osteoconductivity of porous bioactive titanium. Biomaterials **26**, 6014–6023 (2005)
118. P. Habibovic, K. de Groot, Osteoinductive biomaterials–properties and relevance in bone repair. J. Tissue Eng. Regen. Med. **1**, 25–32 (2007)
119. N.K. Tolochko, V.V. Savich, T. Laoui, L. Froyen, G. Onofrio, E. Signorelli, V.I. Titov, Dental root implants produced by the combined selective laser sintering/melting of titanium powders. Proc. Inst. Mech. Eng. Part L: J. Mater. Des. Appl. **216**, 267–270 (2002)
120. W.Y. Yeong, N. Sudarmadji, H.Y. Yu, C.K. Chua, K.F. Leong, S.S. Venkatraman, Y.C. Boey, L.P. Tan, Porous polycaprolactone scaffold for cardiac tissue engineering fabricated by selective laser sintering. Acta Biomaterialia **6**, 2028–2034 (2010)
121. S.F. Yang, K.F. Leong, Z.H. Du, C.K. Chua, The design of scaffolds for use in tissue engineering. Part I. Traditional factors. Tissue Eng. **7**, 679–689 (2001)
122. S.F. Yang, K.F. Leong, Z.H. Du, C.K. Chua, The design of scaffolds for use in tissue engineering. Part II. Rapid prototyping techniques. Tissue Eng. **8**, 1–11 (2002)

123. S.M. Ahmadi, G. Campoli, S. Amin Yavari, B. Sajadi, R. Wauthle, J. Schrooten, H. Wienans, A.A. Zadpoor, Mechanical behavior of regular open-cell porous biomaterials made of diamond lattice unit cells. J. Mech. Behav. Biomed. Mater. **34**, 106–115 (2014)
124. L. Mullen, R.C. Stamp, W.K. Brooks, E. Jones, C.J. Sutcliffe, Selective laser melting: a regular unit cell approach for the manufacture of porous, titanium, bone in-growth constructs, suitable for orthopedic applications. J. Biomed. Mater. Res. Part B: Appl. Biomater. **89**, 325–334 (2009)
125. S. Van Bael, Y.C. Chai, S. Truscello, M. Moesen, G. Kerckhofs, H. Van Oosterwyck, J.P. Kruth, J. Schrooten, The effect of pore geometry on the in vitro biological behavior of human periosteum-derived cells seeded on selective laser-melted Ti6Al4V bone scaffolds. Acta Biomaterialia **8**, 2824–2834 (2012)
126. L. Le Guehennec, M.-A. Lopez-Heredia, B. Enkel, P. Weiss, Y. Amouriq, P. Layrolle, Osteoblastic cell behaviour on different titanium implant surfaces. Acta Biomaterialia **4**, 535–543 (2008)
127. H. Rotaru, G. Armencea, D. Spîrchez, C. Berce, T. Marcu, D. Leordean, S.G. Kim, S.W. Lee, C. Dinu, G. Băciuţ, M. Băciuţ, In vivo behavior of surface modified Ti6Al7Nb alloys used in selective laser melting for custom-made implants. A preliminary study. Rom. J Morphol. Embryol. **54**, 791–796 (2013)
128. M. de Wild, R. Schumacher, K. Mayer, E. Schkommodau, D. Thoma, M. Bredell, A. Kruse Gujer, K.W. Grätz, F.E. Weber, Bone regeneration by the osteoconductivity of porous titanium implants manufactured by selective laser melting: A histological and micro computed tomography study in the rabbit. Tissue Eng.—Part A **19**, 2645–2654 (2013)
129. W. Mróz, B. Bunder, R. Syroka, K. Niedzielski, G. Golański, A. Slósarczyk, D. Schwarze, T.E.L. Douglas, In vivo implantation of porous titanium alloy implants coated with manesium-doped octacalcium phosphate and hydroxyapatite thin films using pulsed laser deposition. J. Biomed. Mater. Res.—Part B Appl. Biomater. **103**, 151–158 (2015)
130. S.L. Sing, W.Y. Yeong, F.E. Wiria, B.Y. Tay, Characterization of titanium lattice structures fabricated by selective laser melting using an adapted compressive test method. Exp. Mech. (2016)
131. M. Smith, Z. Guan, W.J. Cantwell, Finite element modelling of the compressive response of lattice structures manfactured using selective laser melting technique. Int. J. Mech. Sci. **67**, 28–41 (2013)
132. J. Bültmann, S. Merkt, C. Hammer, C. Hinke, U. Prahl, Scalability of the mechanical properties of selective laser melting produced micro-struts J. Laser Appl. **27** (2015). S29206-29201-29207
133. K. Ushijima, W.J. Cantwell, R.A.W. Mines, S. Tsopanos, M. Smith, An investigation into the compressive properties of stainless steel micro-lattice structures. J. Sandw. Struct. Mater. **13**, 303–329 (2010)
134. F. Brenne, T. Niendorf, H.J. Maier, Additive manufactured cellular structures: Impact of microstructure and local strains on the monotonic and cyclic behavior under uniaxial and bending load. J. Mater. Process. Technol. **213**, 1558–1564 (2013)
135. S. Amin Yavari, R. Wauthle, J. van Der Stok, A.C. Riemslag, D. Janssen, M. Mulier, J.P. Kruth, J. Schrooten, H. Weinans, A.A. Zadpoor, Fatigue behavior of porous biomaterials manufactured using selective laser melting. Mater. Sci. Eng. C **33**, 4849–4858 (2013)

Chapter 3
Formation of Titanium-Tantalum Alloy Using Selective Laser Melting

This chapter introduces the β stabilisers for titanium. The powder preparation for selective laser melting (SLM) of titanium-tantalum is detailed. The powder prepared is then characterised and described. The optimising criteria and optimised process parameters for TiTa is also presented in this chapter.

3.1 β Stabilisers for Titanium

From literature review, there have been several studies done on SLM of new titanium based alloys, such as Ti6Al7Nb [1–3], however studies on β titanium have been limited. β titanium alloys provide enhanced properties such as lower modulus, higher corrosion resistance and improved tissue response when compared to the commonly used titanium alloys [4, 5]. Several metallic elements serve as β stabilizers for titanium, for example iron, vanadium, niobium, molybdenum, nickel, chromium, copper and tantalum. Among the β stabilizing elements, tantalum has low cytotoxic and exhibits excellent biocompatibility [6, 7].

In this work, formation of titanium-tantalum alloy is achieved using SLM for the first time.

3.2 Powder Preparation

Both commercially pure titanium (cpTi) and tantalum powders are produced by gas atomization. The cpTi powder (Grade 2 ASTM B348, LPW Technology Ltd) is spherical in shape and has particle size with average 43.5 μm in size. The tantalum powder (Singapore Demand Planner Ltd) is irregular in shape and has average particle size of 44 μm. The morphologies of the cpTi and tantalum powders are shown in Fig. 3.1.

S. L. Sing, *Selective Laser Melting of Novel Titanium-Tantalum Alloy as Orthopaedic Biomaterial*, Springer Theses, https://doi.org/10.1007/978-981-13-2724-7_3

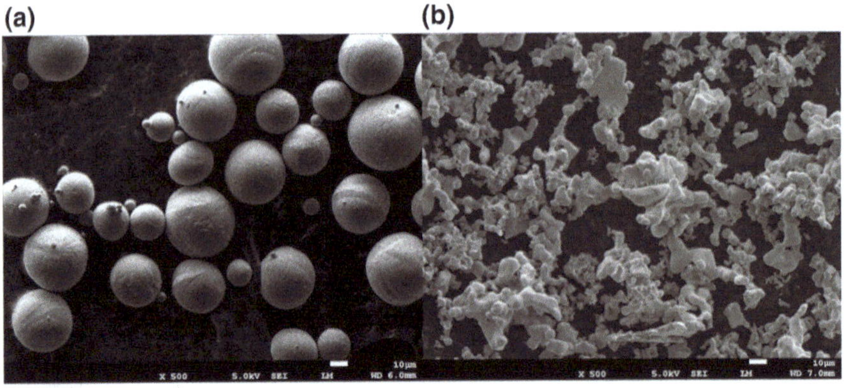

Fig. 3.1 FESEM images of **a** cpTi powder and **b** tantalum powder

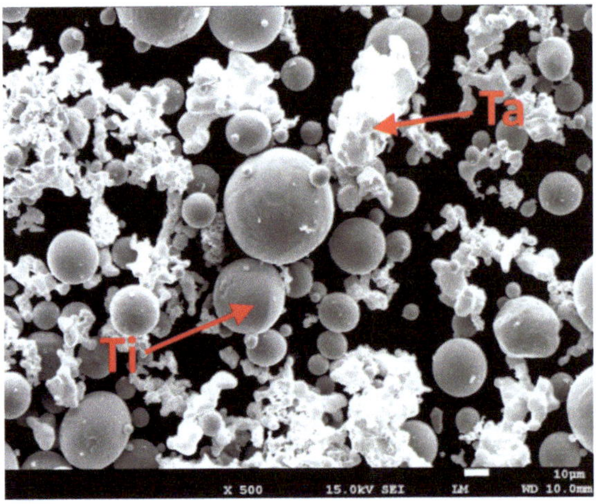

Fig. 3.2 Titanium + tantalum powder mixture. The spherical powder is titanium and the irregular powder is tantalum

The two powders were mixed in weight ratio of 1:1 and then spun at a rate of 60 rpm for 12 h using a tumbler mixer (Inversina 2L, Bioengineering AG). The mixed powder density was measured using gas displacement pycnometry system (AccuPyc II 1340, Micromeritics).

3.3 Powder Characterization

The mixing procedure was effective in mixing these two types of powder together, as shown in Fig. 3.2.

The titanium powder particles remained spherical after the mixing, which is important for the flowability of the powder mixture. Flowability tests were conducted on

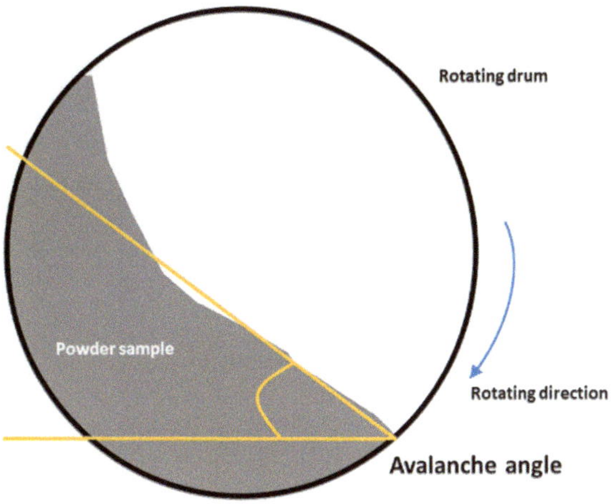

Fig. 3.3 Schematic of the avalanche angle measured in flowability test

the cpTi, tantalum and TiTa powders using the revolution method (Revolution Powder Analyzer, Mercury Scientific Inc.). In this test, a tapped volume of 100 cm^3 of powder was measured by freely filling a cup that was gently tapped until no more powder could fit in. Excess powder was removed using a sharp edge. The powder was then placed inside the cylindrical drum with transparent glass sides. The drum was set to rotate at 0.3 rpm and a digital camera was used to monitor the flow behaviour of the powder. Due to the drum rotation, the powder would be carried up along the side of the drum until it could not support its weight, forming avalanches. The avalanche angle was computed by measuring the angle where the powder was at maximum position before the start of the avalanche. Lower avalanche angles are indicative of better flowability of powders. The avalanche angle is illustrated in Fig. 3.3.

There are advantages in mixing cpTi and tantalum powders, as follows:

(1) Flowability is a key concern in SLM, as uniform powder deposition is required for production of parts with high relative density. Spherical powder is ideally desired. However, the tantalum powder has non-spherical shape, as its high melting point of 2996 °C restricts the production of spherical powder economically. Hence, the overall flowability is improved by mixing the tantalum powder with spherical cpTi powder. The spherical titanium particles roll easier during powder depositions and acts as a medium by pushing the tantalum particles along. Flowability can be indicated by the avalanche angle of the powders. Table 3.1 shows the average avalanche angle of cpTi, tantalum and titanium-tantalum mixed powders for 150 avalanches (n = 150).
    The results show that the TiTa mixed powder has better flowability compared to the non-spherical tantalum powder. However, the blended powder flowability is still lower compared to the cpTi powder. Nonetheless, the improvement in flowability is sufficient for powder deposition and subsequent fabrication by SLM as shown in the experiments conducted in this study.

Table 3.1 Avalanche angle of cpTi, tantalum and TiTa powders

Material	Avalanche angle (deg), n = 150
cpTi	44.65 ±3.19
Tantalum	56.51 ±5.88
TiTa	52.54 ± 4.25

Table 3.2 Physical properties of cpTi and tantalum powders

Materials	Density (g/cm^3)	Melting point (°C)
cpTi	4.51	1668 ±10
Tantalum	16.60	2996

Table 3.3 Chemical composition of TiTa mixed powder

Powder	Nominal (wt%)	ICP-AES (wt%), n = 3	EDS (wt%), n = 5
cpTi	50	51.27	55.74 ±1.06
Tantalum	50	48.19	44.26 ±1.06

Table 3.4 Calculations of theoretical TiTa density

Material	Mass (g)	Density (g/cm^3)	Volume (cm^3)
cpTi	0.5	4.51	0.11
Tantalum	0.5	16.60	0.03
TiTa	1	7.10	0.14

(2) Tantalum has a high density of 16.6 g/cm^3. The powder volume increases for a specific weight of the powder mixture by mixing tantalum powder with cpTi powder, as compared to processing pure tantalum powder, due to the lower density of cpTi (4.51 g/cm^3). This lower the production cost of processing in SLM as larger powder volume allows fabrication of larger parts.

Physical properties of cpTi and tantalum obtained from ASM Handbook are summarized in Table 3.2.

The huge density difference between titanium and tantalum can result in segregation of the powders in the mixture. The TiTa mixed powder composition was ascertained by inductive coupled plasma atomic emission spectroscopy (ICP-AES) and energy dispersive X-ray spectroscopy (EDS). Random samplings were done with the powder mixture. The results are tabulated in Table 3.3, showing no segregation of titanium and tantalum in the powder mixture.

The theoretical density of TiTa is calculated using the density of titanium and tantalum, details of the calculation is shown in Table 3.4.

The theoretical and measured densities of TiTa mixed powder is tabulated in Table 3.5.

Table 3.5 Theoretical and measured densities of TiTa powder

Material	Theoretical density (g/cm^3)	Measured density (g/cm^3), n = 5
TiTa	7.10	7.0835 ±0.0035

3.4 Parameters Optimization for Selective Laser Melting

Fabrication of the TiTa alloy was carried out on a SLM 250 HL machine (SLM Solutions Group AG). The SLM machine is equipped with a Gaussian beam fibre laser with maximum power of 400 W and a focal diameter of 80 μm. All processing occurred in an argon environment with less than 0.05% oxygen to prevent oxidation and degradation of the material during the process [8]. Sectorial, also known as island or chessboard, scanning as shown in Fig. 3.4 was used. This scanning strategy has been reported to minimize thermal stresses formed during the process [9, 10].

To optimize the processing parameters for fabrication of TiTa using SLM, a series of experiments is conducted by varying the hatch spacing and scanning speed while keeping the laser power and layer thickness constant.

The key parameters in SLM are shown in Fig. 3.5.

The energy density E_d of the selective laser melting process is influenced by the laser power P, hatch spacing H, laser scanning speed S and layer thickness L of the powder according to the equation below:

$$E_d = \frac{P}{S.H.L} \tag{3.1}$$

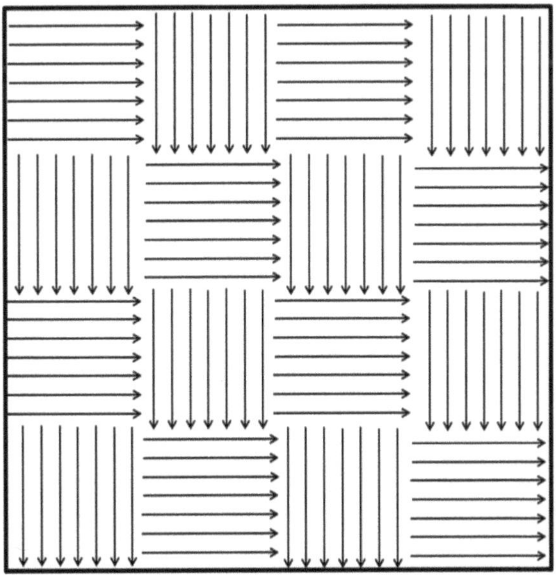

Fig. 3.4 Chessboard strategy used in SLM

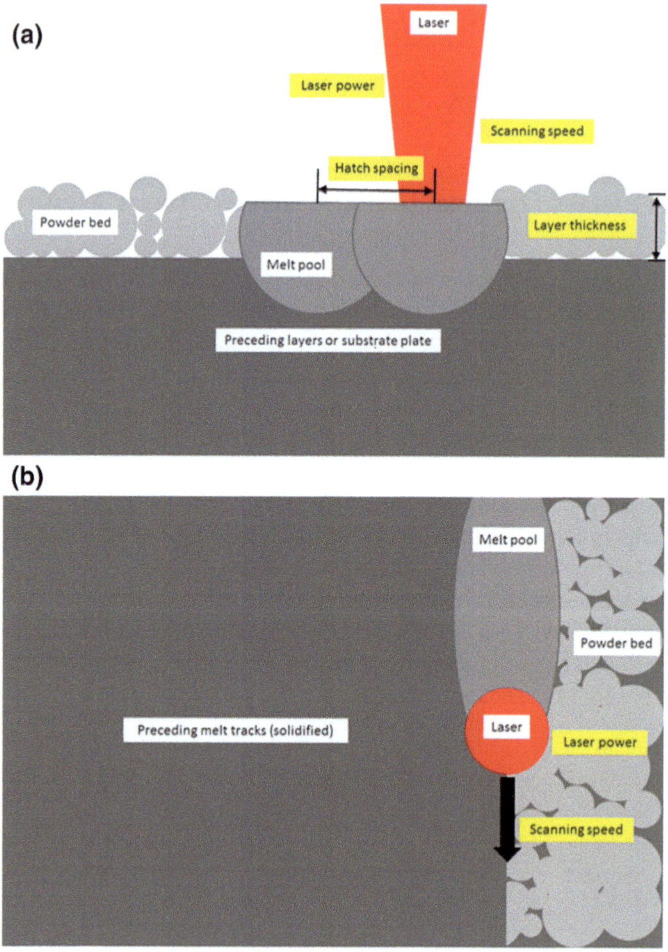

Fig. 3.5 SLM process parameters **a** laser power, hatch spacing, layer thickness **b** scanning speed

The processing parameters are optimized and chosen based on the relative density achieved for the sample and the formed macrostructure of the samples studied under optical microscopy (OM, SZX 7, Olympus). The sample densities were measured using a XS Analytical Balance with sensitivity of 0.001 g and repeatability of 0.0001 g (XS 204, Mettler Toledo), based on the Archimedes Principle. Archimedes measurements are based on a combination of dry weighing and weighing in deionized water. The absolute density of each specimen is calculated using the equation:

$$\rho_{abs} = \frac{m_{air}}{m_{air} - m_{water}} \times \rho_{water} \tag{3.2}$$

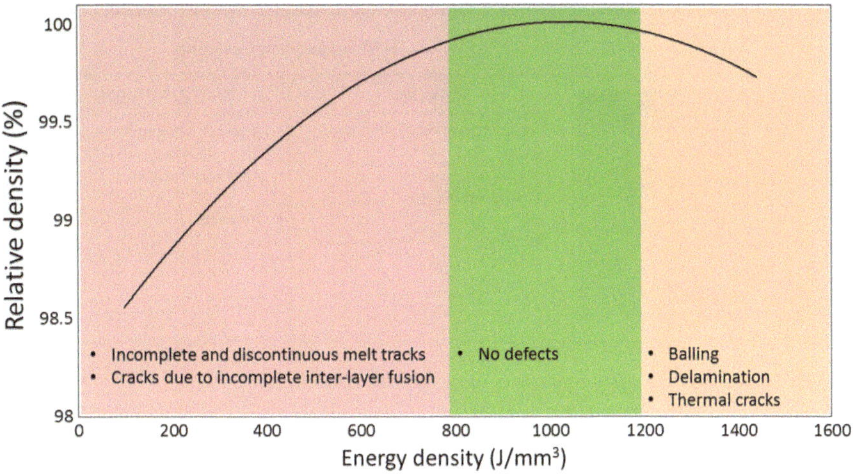

Fig. 3.6 General relationship between relative density of SLM parts and energy density

where ρ_{abs} is the absolute density of the specimen, m_{air} is the mass of specimen in dry weighing, m_{water} is the mass of specimen fully submerged in deionized water and ρ_{water} is the density of deionized water. The relative density $\rho_{relative}$ of the specimen is then calculated using the equation:

$$\rho_{relative} = \frac{\rho_{abs}}{\rho_{theoretical}} \times 100\% \tag{3.3}$$

where the theoretical density $\rho_{theoretical}$ of TiTa is taken to be 7.10 g/cm^3, as shown in Table 3.5.

A general trend for density measurements and macrostructure analysis is graphically represented in Fig. 3.6. Insufficient energy density will result in incomplete and discontinuous melt tracks, as the energy input is not sufficient in melting all the powder materials. The melt pools formed with low energy density also do not have enough penetration depth to allow complete inter-layer fusion, resulting in cracks in the part. Optimal energy density will result in melting of the powder materials to form good fusion between the powder layers, and resulting in parts of near full density. Excessive melting occurs when the energy density input is too high, leading to balling, delamination as a result of thermal cracks due to the large temperature difference between the layers.

SLM is a complex metallurgical process. Thus, optimising the above mentioned inter-related parameters is crucial in producing high quality parts with high relative density. Unfavourable defects may occur in the parts due to localised irregularities such as balling, cracks, delamination and residual stress. Furthermore, unmelted or partially melted segments and porosity may be created due to improper powder deposition which leads to low and inhomogeneous powder bed density, and reduction

Sample	OM Inspection Criteria	
	XY-plane	YZ/XZ-plane
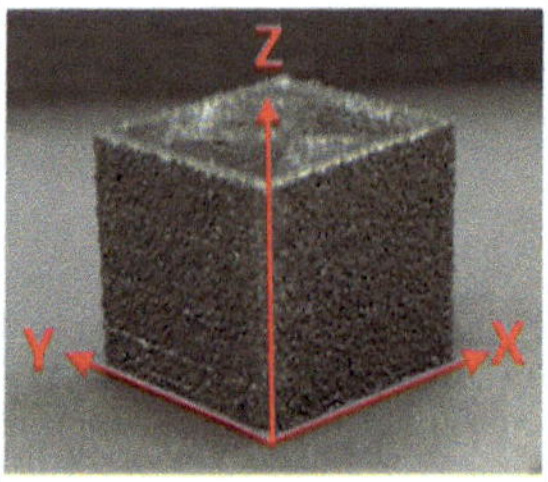	• Complete and continuous melt tracks • Absence of balling • No pores	• Absence of delamination • Absence of cracks • No pores

Fig. 3.7 Sample of TiTa alloy for density measurement and selection criteria

in solubility of some elements in melt during solidification [11, 12]. These defects will lead to undesirable effects on the relative density of the SLM parts.

Balling is the droplet or fragmentation formation from the melt pool as a result of capillary instability [13]. The stability of melt pool is affected by increasing scanning speed which result in elongated liquid pool. During SLM process, the laser scanning causes melting in powder along straight lines, resulting in formation of a continuous cylindrical liquid track. The surface energy of the liquid track may reduce until the breaking of the cylinder to reach equilibrium, leading to metallic fragments of spherical shape to be formed. Balling results in weak bonding between the melt tracks and affects the uniform deposition of the next powder layer. This leads to a compounded effect and may form porosity. The weak interlayer bonding can cause delamination and thermal stresses [13]. Balling can be avoided by improving the stability of the melt pool by careful control of the process parameters.

Cracks significantly diminish the mechanical properties of SLM parts. Cracks in SLM parts can be classified into microscopic and macroscopic cracks. The microscopic cracks are usually formed during rapid solidification due to liquid film interruption at grain boundaries as a result of tensile stress [13, 14]. Macroscopic cracks are formed due to low ductility of the material and stress induced crack propagation.

An example of the specimen used for density measurement and OM with the inspection criteria are shown in Fig. 3.7.

The samples are studied under OM to ascertain the macrostructure of the samples of various combinations of processing parameters. The processing parameters used and the examples of resulting surfaces in the xy-plane are shown in Fig. 3.8.

Selected defects, such as balling and interlayer cracks, observed in some samples are highlighted in Figs. 3.9 and 3.10.

When energy density is high, excess energy in the laser beam leads to the Marangoni effect, which in turn affects the quality of the melt pool [15, 16]. The Marangoni effect is the mass transfer between melt pools along their interface due to surface tension gradient. Due to the thermo-capillary action, Marangoni effect can affect the penetration depth of the melt pool [17]. Due to the influence of Marangoni

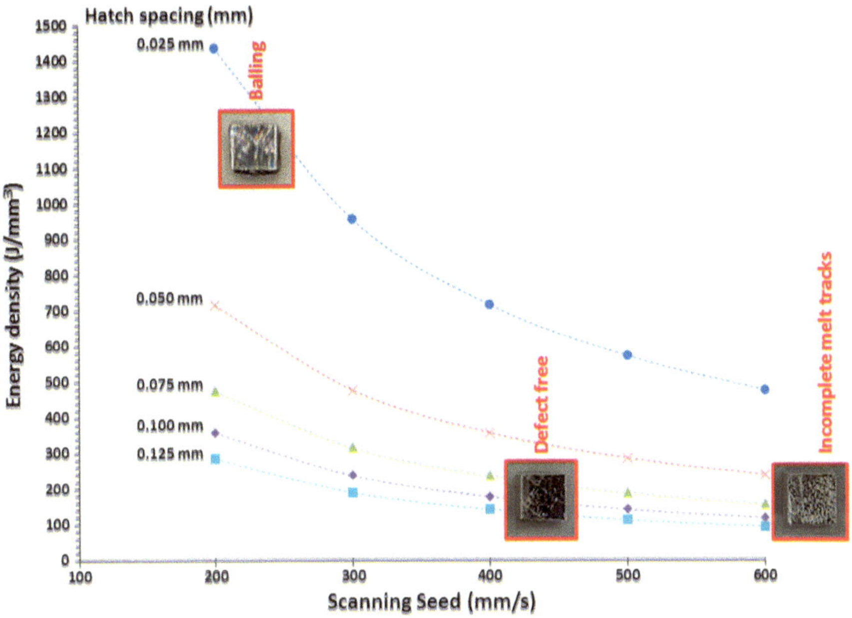

Fig. 3.8 xy-plane surface morphology of density specimens with variation in energy density input

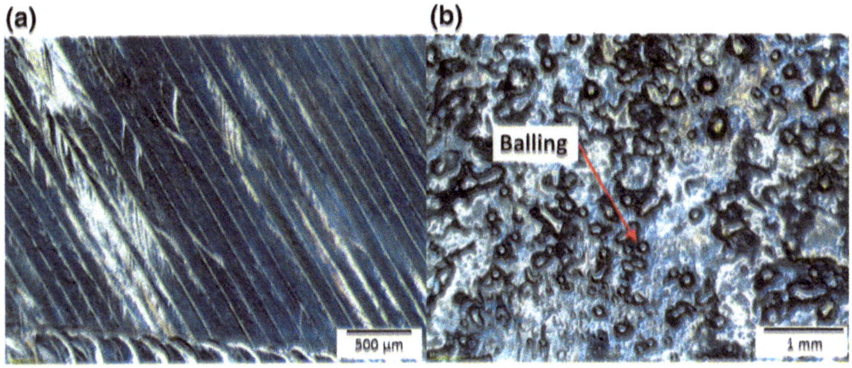

Fig. 3.9 OM images showing **a** complete melt tracks **b** balling

effect, it can lead to instability of flow in the melt pool. The instability of flow in a melt pool leads to the breaking up of thin melt pools into spherical droplets, resulting in balling.

Interlayer cracks are often caused by formation of small melt pools with low penetration depth resulting in incomplete fusion across the layers as well as build-up of thermal gradients across layers resulting in delamination. Furthermore, shear stress formed in the liquid phase during the melting process can cause increased surface tension of the melt pool which leads to melt pool instability and cracks. Due

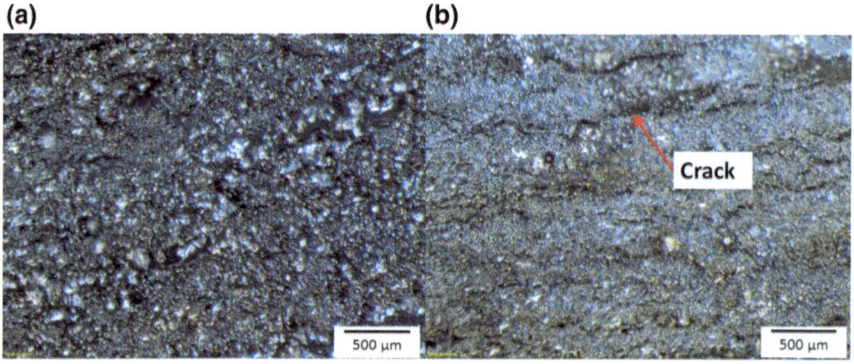

Fig. 3.10 OM images showing **a** complete fusion across layers **b** interlayer cracks

to the rapid heating and cooling in SLM, residual stress may be formed. The residual stress can result in stress cracking and interlayer delamination. It is reported that residual stress usually include large tensile stress at the top and bottom of the SLM parts and compression stress at the segment in between. The molten top layers tend to shrink due to thermal contraction, which is restricted by the previous solidified layers. Thus, tensile stress is induced in the added top layers. When successive layers are added on top, each layer induces a certain stress profile in the substrate and also in the underlying solidified layers, thus reducing the initial tensile stress present in these layers, resulting in compression stress at the segment in between [18]. The residual stress is dependent on the material properties such as elastic modulus and thermal expansion coefficient, sample height and processing parameters. In general, residual stress is larger in the direction along the build direction. The residual stress can be controlled by preheating the build substrate and reducing the temperature gradient [19].

3.5 Summary

The processing window of TiTa by SLM which achieved 99.9%, full dense parts is tabulated in Table 3.6. However, some of the full dense parts are not defect free.

Table 3.6 SLM processing window for TiTa

Process parameters	
Laser power (W)	360
Laser scan speed (mm/s)	200–600
Layer thickness (µm)	50
Hatch spacing (mm)	0.025–0.125

References

1. E. Chlebus, B. Kuznicka, T. Kurzynowski, B. Dybala, Microstructure and mechanical behaviour of Ti-6Al-7Nb alloy produced by selective laser melting. Mater. Charact. **62**, 488–495 (2011)
2. T. Marcu, M. Todea, L. Maines, D. Leordean, P. Berce, C. Popa, Metallurgical and mechanical characterisation of titanium based materials for endosseous applications obtained by selective laser melting. Powder Metall. **55**, 309–314 (2012)
3. P. Szymczyk, A. Junka, G. Ziolkowski, D. Smutnicka, M. Bartoszewicz, E. Chlebus, The ability of S.aureus to form biofilm on the Ti-6Al-7Nb scaffolds produced by Selective Laser Melting and subjected to the different typs of surface modifications. Acta Bioeng. Biomech. **15,** 69–76 (2013)
4. D. Kuroda, M. Niinomi, M. Morinaga, Y. Kato, T. Yashiro, Design and mechanical properties of new β type titanium alloys for implant materials. Mater. Sci. Eng., A **243**, 244–249 (1998)
5. M. Niinomi, D. Kuroda, K.-I. Fukunaga, M. Morinaga, Y. Kato, T. Yashiro, A. Suzuki, Corrosion wear fracture of new β type biomedical titanium alloys. Mater. Sci. Eng., A **263**, 193–199 (1999)
6. Y. Okazaki, Y. Ito, K. Kyo, T. Tateishi, Corrosion resistance and corrosion fatigue strength of new titanium alloys for medical implants without V and Al. Mater. Sci. Eng., A **213**, 138–147 (1996)
7. E. Eisenbarth, D. Velten, M. Müller, R. Thull, J. Breme, Biocompatibility of β-stabilizing elements of titanium alloys. Biomaterials **25**, 5705–5713 (2004)
8. L.E. Loh, Z.H. Liu, D.Q. Zhang, M. Mapar, S.L. Sing, C.K. Chua, W.Y. Yeong, Selective laser melting of aluminium alloy using a uniform beam profile. Virtual Phys. Prototyp. **9**, 11–16 (2014)
9. J. Jhabvala, E. Boillat, T. Antignac, R. Glardon, On the effect of scanning strategies in the selective laser melting process. Virtual Phys. Prototyp. **5**, 99–109 (2010)
10. E. Yasa, J. Deckers, J.-P. Kruth, M. Rombouts, J. Luyten, Investigation of sectoral scanning in selective laser melting, in: *ASME 2010 10th Biennial Conference on Engineering Systems Design and Analysis, Istanbul, Turkey*, 2010
11. J.P. Kruth, G. Levy, F. Klocke, T.H.C. Childs, Consolidation phenomena in laser and powder-bed based layered manufacturing. CIRP Ann. Manuf. Technol. **56**, 730–759 (2007)
12. K.G. Prashanth, S. Scudino, H.J. Klauss, K.B. Surreddi, L. Löber, Z. Wang, A.K. Chaubey, U. Kühn, J. Eckert, Microstructure and mechanical properties of Al-12Si produced by selective laser melting: Effect of heat treatment. Mater. Sci. Eng., A **590**, 153–160 (2014)
13. D.D. Gu, W. Meiners, K. Wissenbach, R. Poprawe, Laser additive manufacturing of metallic components: materials, processes and mechanisms. Int. Mater. Rev. **57**, 133–164 (2012)
14. M. Zhong, H. Sun, W. Liu, X. Zhu, J. He, Boundary liquation and interface cracking characterization in laser deposition of Inconel 738 on directionally solidified Ni-based superalloy. Scripta Mater. **53**, 159–164 (2005)
15. N.K. Tolochko, S.E. Mozzharov, I.A. Yadroitsev, T. Laoui, L. Froyen, V.I. Titov, M.B. Ignatiev, Balling processes during selective laser treatment of powders. Rapid Prototyp. J. **10**, 78–87 (2004)
16. J.-P. Kruth, P. Mercelis, J. Van Varenbergh, L. Froyen, M. Rombouts, Binding mechanisms in selective laser sintering and selective laser melting. Rapid Prototyp. J. **11**, 26–36 (2005)
17. B. Xiao, Y. Zhang, Numerical simulation of direct metal laser sintering of single-component powder on top of sintered layers. J. Manuf. Sci. Eng. **130**, 041002 (2008)
18. P. Mercelis, J.P. Kruth, Residual stresses in selective laser sintering and selective laser melting. Rapid Prototyp. J. **12**, 254–265 (2006)
19. H. Yves-Christian, W. Jan, M. Wilhelm, W. Konrad, P. Reinhart, Net shaped high performance oxide ceramic parts by selective laser melting. Phys. Procedia **5**, 587–594 (2010)

Chapter 4
Characterisation of Selective Laser Melted Titanium-Tantalum Alloy

This chapter details the microstructural and mechanical characterisation of titanium-tantalum (TiTa) alloy formed by selective laser melting (SLM). The test methods for metallographic and mechanical characterisations are documented. The results are then discussed and analysed using the theories behind the SLM process. Benchmarking against commonly used commercially pure titanium (cpTi) and Ti6Al4V is also done for the TiTa alloy.

This set of optimised parameters, as tabulated in Table 4.1, are chosen to fabricate all the TiTa alloy bulk samples based on the relative density and macrostructure analysis results using optical microscopy (OM), as shown in Chap. 3. The optimised parameters are chosen as the specimens produced have relative density of 99.9%, full dense, and are free from defects.

4.1 Metallographic Analysis of Titanium-Tantalum Parts

The SLM samples were subjected to standard metallographic procedure which is grinding with 320, 800 and 1200 SiC papers in a consecutive manner and then polished by diamond suspensions of 9, 3 and 1 μm. The samples were then etched with Kroll's reagent (10 mL of HF, 30 mL of HNO_3 and 50 mL of water) from ASTM E407 (Standard Practice for Microetching Metals and Alloys) for 20 s. The microstructure

Table 4.1 Optimised SLM processing parameters for TiTa

Process parameters	
Laser power (W)	360
Laser scan speed (mm/s)	400
Layer thickness (μm)	50
Hatch spacing (mm)	0.125

S. L. Sing, *Selective Laser Melting of Novel Titanium-Tantalum Alloy as Orthopaedic Biomaterial*, Springer Theses, https://doi.org/10.1007/978-981-13-2724-7_4

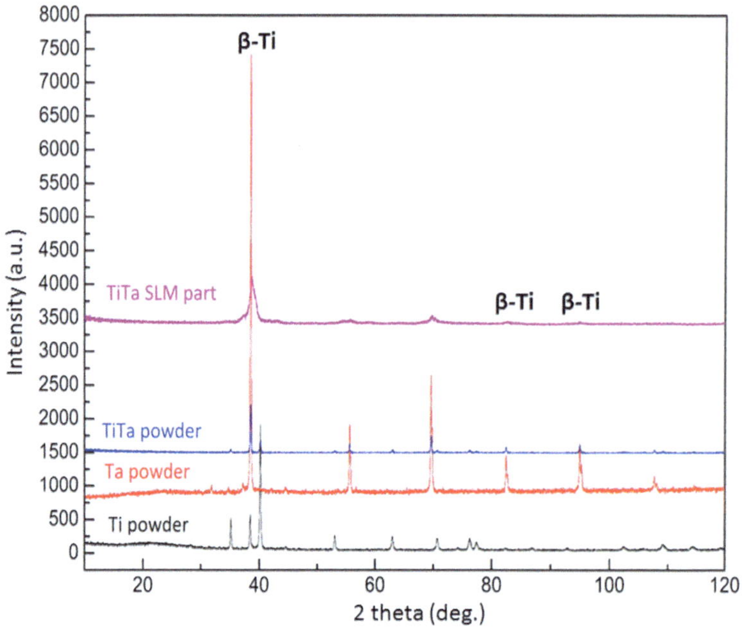

Fig. 4.1 XRD patterns of cpTi, tantalum, TiTa powders and SLM produced TiTa

study was conducted using field emission scanning electron microscopy (FESEM, JSM7600, JEOL), X-ray diffraction (XRD, Empyrean, PANalytical) and electron back scattered diffraction (EBSD, Nordlys, Oxford Instruments). The grain size distribution was measured from the EBSD results using post-processing software, HKL CHANNEL5 (Oxford Instruments).

4.1.1 X-Ray Diffraction Phase Analysis

It is observed from Fig. 4.1 that the peaks in the TiTa mixed powder pattern correspond to the respective peaks of the tantalum powder and cpTi powder. This implies that the physical mixing of the powders did not result in any phase change in the materials.

The XRD spectrums of SLM produced TiTa samples are also shown in Fig. 4.1. After the SLM process, the respective diffraction peaks of titanium and tantalum can still be observed. However, only the peaks corresponding to the β phase are observed. Pure titanium has a hexagonal close packed (HCP) structure, i.e. α phase, at ambient temperature. At temperature greater than 883 °C, it exists as a body centered cubic (BCC) structure, i.e. β phase. The β phase becomes stable at temperatures lower than 883 °C when β stabilizers are added and can be maintained in the metastable state at ambient temperature. The BCC structure stability depends on the extent of alloying elements. The amount of β stabilizer required to retain purely β phase at ambient temperature depends on the Molybdenum equivalency [1], an empirical rule derived from analysis of binary titanium alloys. In general, approximately 10 wt% of molybdenum is required to stabilize the β phase during quenching [2]. During SLM, the parts undergo rapid cooling which is similar to rapid quenching. Molybdenum equivalence is given by:

$$Mo_{eq} = 1.0\ \mathrm{Mo} + 0.67\ \mathrm{V} + 0.44\ \mathrm{W} + 0.28\ \mathrm{Nb} + 0.22\ \mathrm{Ta} + 1.6\ \mathrm{Cr} + \cdots - 1.0\ \mathrm{Al} \tag{4.1}$$

Based on the Molybdenum equivalence, the TiTa alloy formed has a $\mathrm{Mo_{eq}}$ of 11, which is more than 10 but less than the critical value of 25. It signifies that the resulting β titanium from SLM of the mixture of cpTi and tantalum powders is metastable. β titanium and tantalum have similar atomic radii (approximately 0.2 nm) and both have BCC structures with lattice parameter of approximately 332 pm, their XRD peaks coincide. They also share the same peaks as β(Ti, Ta) solid solution [3].

The addition of tantalum in the alloy suppresses the transformation of β phase to the α' phase due to the β stabilizing effect. This was achieved by decreasing the critical cooling rate to retain β phase and lowering of the martensitic start temperature. Coupled with the rapid solidification during SLM process, SLM produced TiTa exhibits single β phase microstructure and not α + β despite being metastable. Previous studies have also shown the preference of formation and growth of β phase over α phase at large undercooling [4]. Metastable β titanium alloys are advantageous as their mechanical properties can be tailored [5]. This implies that the SLM produced TiTa parts can be heat treated to obtain various combinations of mechanical properties for different applications.

4.1.2 Microstructure

The resulting microstructures of the SLM samples in xy-plane and yz-plane are shown in Fig. 4.2(a) and Fig. 4.2(b) respectively.

The SLM produced samples consist of TiTa solid solution matrix with unmelted tantalum particles. The applied energy density during SLM is sufficient to fully melt the titanium powder but some of the larger tantalum particles only melted partially due to the higher melting point of tantalum. The composition of the TiTa matrix was determined to be 50.74 ± 0.82 wt% of titanium and 49.26 ± 0.82 wt% of tantalum. The composition of the TiTa matrix is consistent throughout, even near the boundary of the unmelted tantalum. The consistency of the composition shows that the diffusion of melted tantalum into the matrix is not obstructed by the partial melting. The relatively large two phase (liquid + solid) field in the binary titanium-tantalum phase diagram also shows the difficulty in melting the two materials together. This resulted in the tantalum particles in TiTa matrix microstructure shown in Fig. 4.2.

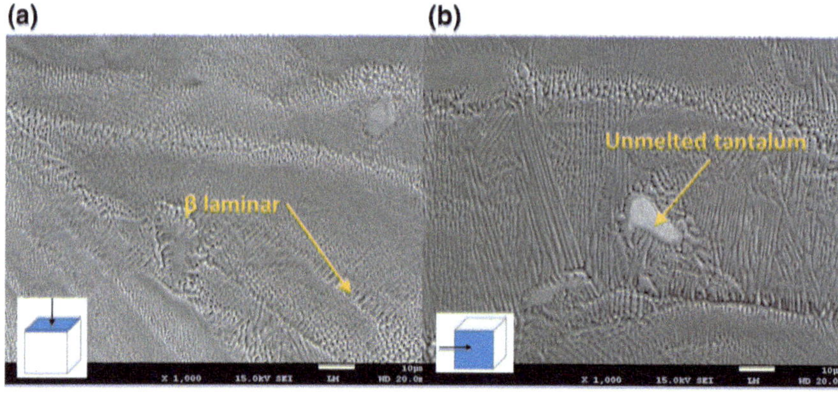

Fig. 4.2 FESEM micrograph of SLM produced TiTa samples **a** xy- and **b** yz-plane

Equiaxed β grains can be observed, as shown in Fig. 4.2. Due to the isomorphous β stability effect of tantalum, no α phase is formed during the rapid cooling in SLM. The equiaxed β grains arise due to the melting of scan tracks and layers that resulted in temperature about the β transus. Partial remelting between adjacent scan tracks resulted in melt tracks formation that is larger than the laser spot size of 80 μm. Furthermore, there was also remelting of previous layer due to penetration depth of the laser larger than the layer thickness of 50 μm. This allows growth of the grains that are parallel, inclined or perpendicular to the build direction in various sizes. This indicates that the grains grow from multiple locations in the melt pool. A schematic showing the overlapping of melt tracks and melt pools is shown in Fig. 4.3.

In addition, the main driving force in the molten pools of SLM process is the convection applied by the combination of surface tension gradient, viscous shear stress and buoyancy forces [6]. During solidification of the molten pool in SLM, the laser beam moves forward and thermal energy is quickly dissipated to the substrate or previous solidified layer due to the higher thermal conductivity of solid compared to the surrounding powder. This results in temperature gradients within the melt pool. The temperature gradients result in chemical potential gradient of solute elements and different directions of liquid flow. These phenomena result in random orientation of the grains formed due to multiple mass flow directions. Furthermore, during SLM, laser scanning is performed line-by-line, followed by layer-by-layer. These influenced the grain formation in multiple directions, as observed in Fig. 4.2, due to corresponding thermal flows.

The directions of the resulting grains are determined using EBSD, as shown in Fig. 4.4.

Like the FESEM images, the EBSD images shows equiaxed β grains developed and grew within each layer in multiple directions. The random grain orientations result in anisotropic microstructure without any preferred grain orientation, despite the rapid solidification rate. Furthermore, due to the scanning strategy which involves

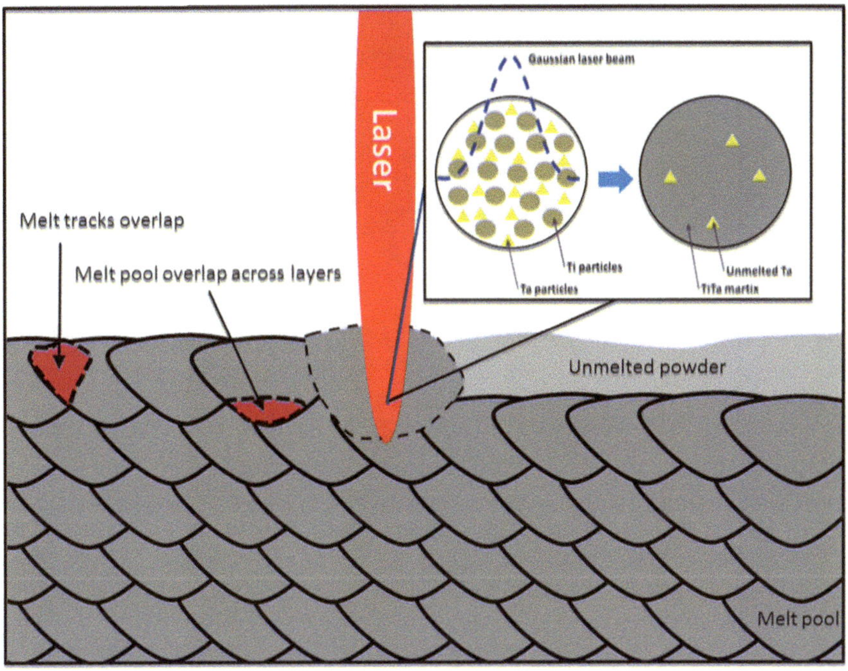

Fig. 4.3 Schematic of overlapping melt tracks and melt pools during SLM

the laser beam moving backwards and forwards, the grain structures produced consist of grains in random orientations.

The grain size in SLM produced TiTa samples ($n=5$) has an average value of 10.20 ± 7.68 μm and the grain size distribution is shown in Fig. 4.5. It is observed that the grain size of the samples is generally smaller than the laser spot size of 80 μm.

This observation implied that there were multiple nucleus sites for grain formation along one single melt track. The grains formed were orientated in various orientations due to the temperature gradients that existed in multiple orientations.

The difference in grain size can be attributed to the difference in thermal conductivity of titanium (21.9 W/mK) and tantalum (57.5 W/mK). When the grain nucleates in proximity to the unmelted tantalum, heat is conducted away quicker due to the higher thermal conductivity of tantalum. This results in smaller grains formed as the solidification rate is higher.

In solidification process, the temperature gradient in the liquid phase, G and the growth of the interface, or solidification rate, R are the two main factors that affect the grain morphology. From literature, a high G/R ratio will result in columnar growth, while low G/R ratio result in equiaxed growth [7]. From the titanium-tantalum phase diagram, the temperature range between liquidus and solidus is approximately 300 °C for titanium with 50 wt% tantalum which allows for solidification to occur within

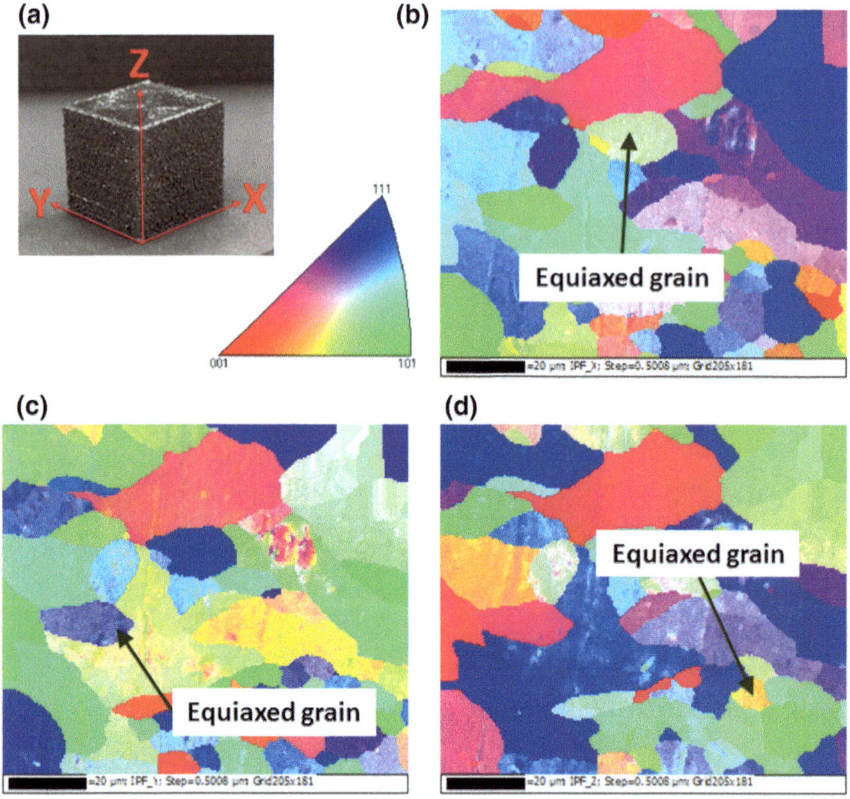

Fig. 4.4 EBSD maps showing different orientation of grains **a** specimen and grain orientation **b** xy- **c** yz- and **d** xz-planes with respect to build orientation

short time periods during SLM. Furthermore, rapid solidification occurs during SLM inherently, hence, R is high during SLM. The sectorial scan strategy used in this study has shorter scan tracks which tend to produce lower temperature gradients (low G) [8]. The low G/R ratio resulting in formation of the equiaxed grains in SLM produced TiTa.

Within each of the equiaxed grains, laminar β phase substructures with directionality can be observed. The substructure surrounds the unmelted tantalum particles grow in direction parallel to the surface of the particles due to tantalum having higher thermal conductivity compared to the titanium-tantalum solid solution, which results in higher G surrounding the tantalum particles as heat is dissipated away from the liquid phase faster along the tantalum particles. When the melt tracks overlap, G is also increased due to remelting. This results in formation of laminar substructures in the solidified TiTa. The formation of these laminar substructure with directionality is due to the Gaussian laser heat source which non uniform power distribution and fluctuating energy output [9]. This results in multiple temperature gradients of

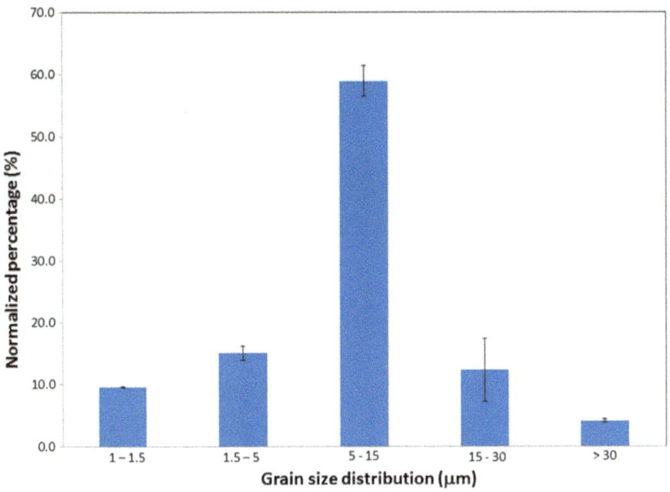

Fig. 4.5 Grain size distribution of SLM produced TiTa samples

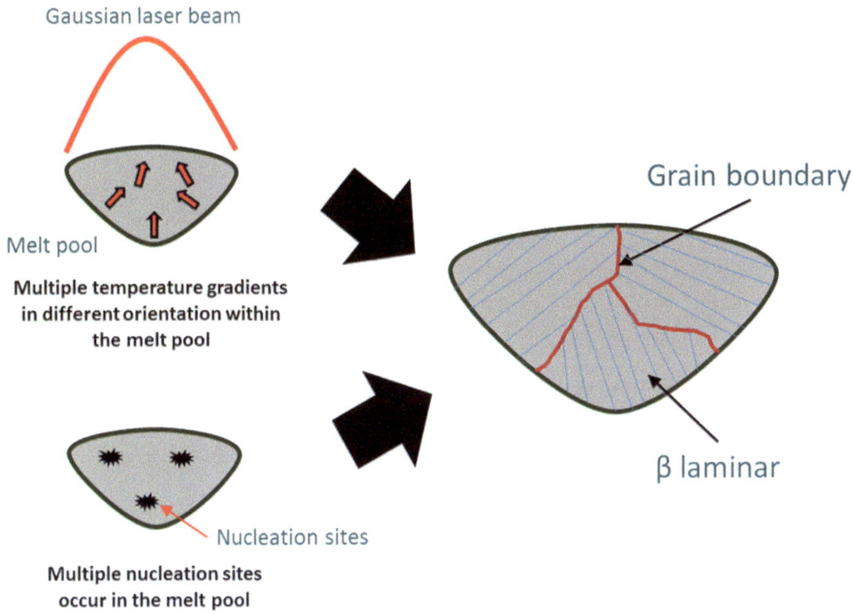

Fig. 4.6 Formation of β laminar substructure in equiaxed grains in SLM produced TiTa

different direction within the melt pool. Coupled with the multiple nucleation sites in the melt pool and differences in thermal conductivity between the liquid, solid and powder, directionality in the substructures occurs.

A schematic showing the formation of laminar substructures is shown in Fig. 4.6.

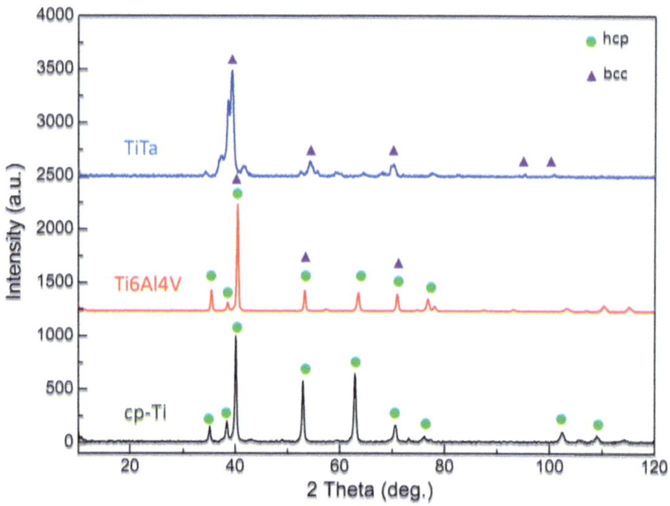

Fig. 4.7 XRD patterns of SLM produced cpTi, Ti6Al4V and TiTa

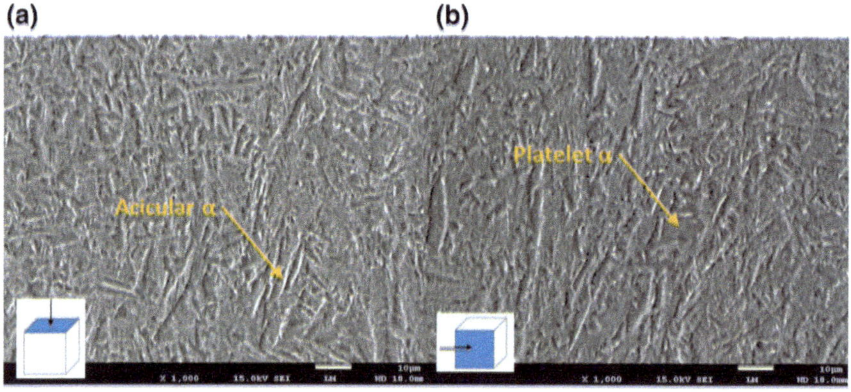

Fig. 4.8 FESEM micrograph of SLM produced cpTi samples **a** xy- and **b** yz-plane

In comparison, SLM produced cpTi consist of primary α phase while SLM produced Ti6Al4V samples consist of α' phase. The phase transformation in the SLM produced materials is determined by solidification behaviour of the molten pools with complete liquid formation, including the liquid flow, solidification rate and thermal history [10]. Heating and cooling over the beta transus temperature leads to complete re-nucleation of phases.

The difference in microstructure between these three materials is captured by XRD analysis as shown in Fig. 4.7.

The resulting microstructures of cpTi and Ti6Al4V samples are also shown in Fig. 4.8 and Fig. 4.9 respectively.

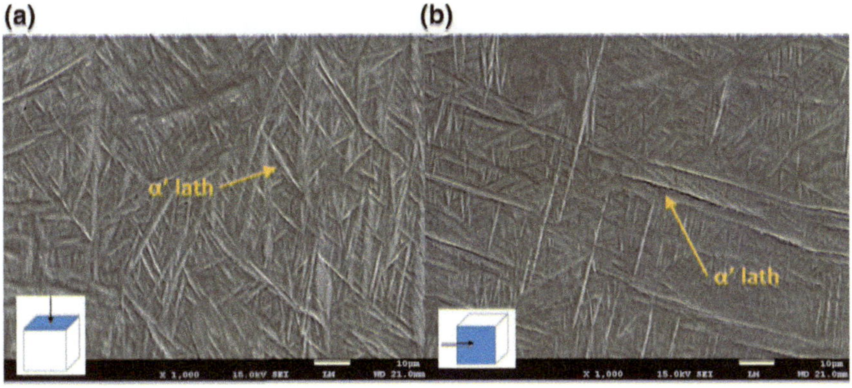

Fig. 4.9 FESEM micrograph of SLM produced Ti6Al4V samples **a** xy- and **b** yz-plane

Standard XRD peaks corresponding to HCP structure were detected for the cpTi and Ti6Al4V samples. The HCP structure is determined to be α phase for cpTi samples from the FESEM images which shows the microstructure consisting of a mixture of acicular and platelet α. The platelet α is formed when rapid cooling occurred during SLM from temperature above the beta transus of 883 °C, allowing the transformation of β into α phase. However, in Ti6Al4V samples, the HCP structure is determined to be α' phase from the FESEM images which shows complete martensitic microstructure. Martensitic laths transformed from prior β grain boundaries and fill the grains. Furthermore, the XRD pattern of Ti6Al4V samples indicates presence of a HCP phase with lattice parameters a = 0.2944 nm and c = 0.4678 which is in agreement with the lattice parameter values given for the α' phase (a = 0.29313 nm and c = 0.46813 nm) in literature [11].

The materials in SLM undergo very high cooling rates during the process. However, these cooling rates vary due to the differences in physical properties of the materials. In addition, the difference in composition of cpTi, Ti6Al4V (with α and β stabilizers) and TiTa (with β stabilizer) results in various beta transus. The difference in cooling rates and beta transus lead to different microstructure formation in the three materials.

4.2 Mechanical Analysis of Titanium-Tantalum Parts

4.2.1 Tensile Properties

Tensile coupons with gauge length of 25 mm, based on ASTM E8 (Standard Test Methods for Tension Testing of Metallic Materials) were produced using electrical

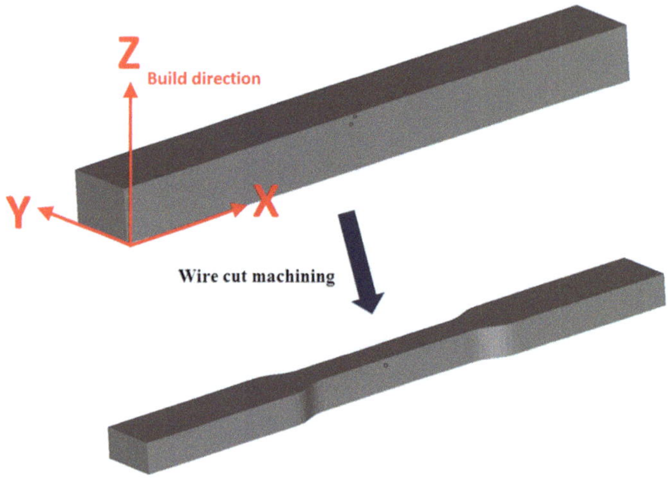

Fig. 4.10 Schematic of producing tensile coupons from blocks fabricated by SLM

Fig. 4.11 Build orientation of blocks fabricated using SLM

discharge machining (EDM) wire cut from SLM fabricated blocks, as shown in Fig. 4.10. This is to ensure that the test coupons used follow the standard.

Tensile test (Static Tester Series 5569, Instron) was conducted with a 50 kN load cell and strain rate of 1 mm/min. Tensile test loading direction was perpendicular to the build direction, i.e. along the xy-plane.

A schematic of build orientation during the SLM process is shown in Fig. 4.11.

The build direction is along the z-axis for all test samples.

The tensile properties of SLM produced cpTi, Ti6Al4V and TiTa are shown in Table 4.2. All tensile coupons were fabricated using parameters stated in Table 4.1.

The corresponding typical stress-strain curves are plotted in Fig. 4.12.

The Young's modulus of SLM produced TiTa is the lowest. The elastic modulus of an alloy is mainly determined by the modulus and volume fractions of the constitution phases and is not sensitive to grain size [12]. It was reported that β phase has the

Table 4.2 Tensile properties of SLM produced TiTa, Ti6Al4V and cpTi samples (n = 5)

Material	Young's modulus (GPa)	Ultimate tensile strength (MPa)	Yield strength (MPa)	Elongation (%)
TiTa	75.77 ± 4.04	924.64 ± 9.06	882.77 ± 19.60	11.72 ± 1.13
Ti6Al4V	131.51 ± 16.40	1165.69 ± 107.25	1055.59 ± 63.63	6.10 ± 2.57
cpTi	111.59 ± 2.65	703.05 ± 16.22	619.57 ± 20.25	5.19 ± 0.32

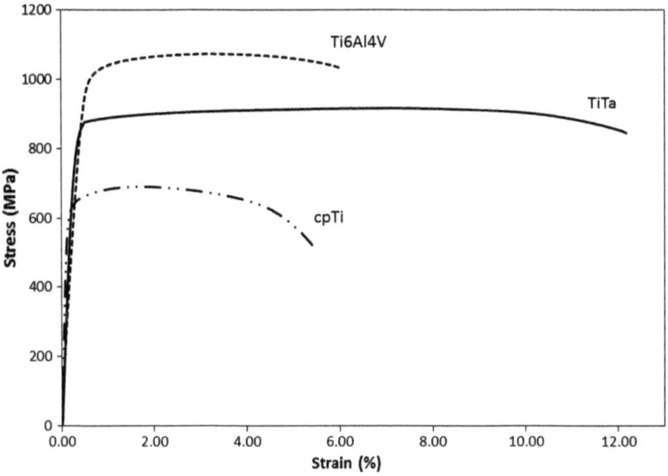

Fig. 4.12 Stress-strain curves of SLM produced TiTa, cpTi and Ti6Al4V specimens

lowest Young's modulus in titanium phases, α' phase has modulus lower than α phase and the phase with the highest modulus is ω phase [13]. The observation from the SLM produced titanium and its alloys parts are in agreement with the reported results. TiTa samples have the lowest Young's modulus, as only β phase is present, followed by cpTi (with α phase) samples. Ti6Al4V (with α' + β phase) samples have the highest Young's modulus. This observation is also in agreement with findings from other β titanium alloys [14]. In addition, the TiTa specimens have higher ductility, shown by the higher elongation at yield, than Ti6Al4V. However, the higher ductility is a trade-off for lower yield strength, which is sensitive to size and morphology of the microstructures [12]. Higher ductility is due to the absence of strain hardening in TiTa alloy. Strain hardening may occur in Ti6Al4V due to the presence of α' martensitic phase [15], which does not exist in the TiTa alloy. Furthermore, the increase in elongation can be attributed to a microstructural change to the more ductile β phase in TiTa alloy [1].

The fracture surfaces after tensile tests for the materials were investigated using FESEM, as shown in Fig. 4.13.

All the specimens fractured after neck creation. The fracture surfaces confirmed the ductility of the three materials. At microscopic level, they exhibited features showing mixed mode of ductile and brittle fracture. The fracture surfaces showed

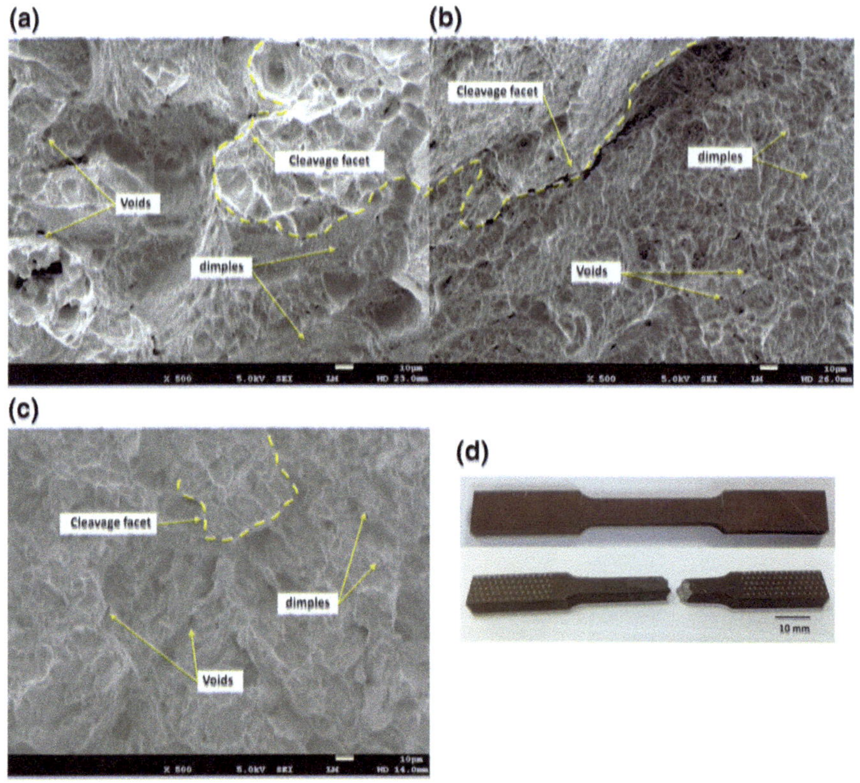

Fig. 4.13 Fracture surfaces after tensile test **a** TiTa **b** Ti6Al4V **c** cpTi samples and **d** samples of tensile test coupon before and after fracture

a combination of ductile dimples and voids indicating ductile failures as well as cleavage facets that consisted of flat planes with small atomic steps indicating brittle fracture. The fractures were proceeded predominantly by ductile intragranular fracture mode with dimple-like morphology. As can be seen from Fig. 4.13a the fracture surface of TiTa is covered by very fine dimples of about 2–5 μm, shown by the brighter lines in the FESEM image. This was compared to the Ti6Al4V which shows larger dimples of about 5–10 μm. The dimple size is indicative of the fracture energy during the fracture. Ductile fracture was more dominant in TiTa compared to cpTi as observed from the smaller cleavage facets and planes on the fracture surface of cpTi. This could be attributed to the presence of the β phase that is more ductile than the α phase. The higher amount of deep and fine ductile dimples in fracture surface of the TiTa specimen also represents higher ductility of the TiTa alloy.

In addition, the mechanical properties of TiTa obtained from SLM is compared to the alloy with same composition obtained by arc melting in previous works conducted by Zhou et al. [16, 17], and the mechanical properties are summarized in Table 4.3.

Table 4.3 Comparison of properties of titanium-tantalum alloy obtained by SLM and arc melting

Method	Phase present	Young's modulus (GPa)	Ultimate tensile strength (MPa)	Yield strength (MPa)	Elongation (%)
SLM (n = 5)	β	75.77 ± 4.04	924.64 ± 9.06	882.77 ± 19.60	11.72 ± 1.13
Arc melting [16, 17]	α"	88	530	375	25

Table 4.4 Micro-hardness of SLM produced TiTa, Ti6Al4V and cpTi samples

Material	xy-plane (HV), n = 10	yz-plane (HV), n = 10
TiTa	284.5 ± 11.06	282.7 ± 9.76
Ti6Al4V	383.16 ± 10.62	386.83 ± 8.73
cpTi	213.4 ± 10.29	217.4 ± 3.67

TiTa alloy obtained from SLM is able to achieve a higher strength to modulus ratio compared to arc melting. This is due to the rapid solidification and cooling that occurs during the SLM process which retained the β phase. The rapid solidification and cooling also result in smaller grains in microstructure which leads to higher strengths.

4.2.2 *Micro-Hardness*

The micro-hardness test of the material was carried out using Vickers hardness test (DuraScan, Struers) on the xy-plane and yz-plane with a load of 100 g and a loading time of 15 s.

The differences in microstructure between the materials also translate to differences in micro-hardness values as shown in Table 4.4.

There is insignificant difference between the micro-hardness values for the xy-plane and yz-plane for TiTa, Ti6Al4V and cpTi respectively. The random grain orientations without any preferred grain orientation due to the scanning strategy which involves the laser beam moving backwards and forwards, results in similar micro-hardness values in the two planes.

4.3 Summary

The capability of SLM to fabricate TiTa alloy was demonstrated for the first time in this chapter. Tantalum was selected as a potential alloying element based on its capability to stabilize the β phase in the TiTa alloy as well as lowering the Young's modulus. A lower Young's modulus is desirable to reduce stress shielding in biomedical implants. Based on specific applications, the alloy content can be altered to reduce the tantalum particles content or heat treatment can be done to induce the required microstructures and mechanical properties.

The key findings include:

(1) TiTa alloy could be fabricated successfully using SLM, demonstrating the SLM capability to process powder mixtures of different materials apart from pre-alloyed powders.
(2) TiTa alloy processed by SLM was shown to consist only of β phase due to tantalum stabilizing effect of the phase after rapid solidification.
(3) TiTa part showed a combination of high strength and lower Young's modulus as compared to cpTi and Ti6Al4V parts.

References

1. B. Vrancken, L. Thijs, J.-P. Kruth, J. Van Humbeeck, Microstructure and mechanical properties of a novel β titanium metallic composite by selective laser melting. Acta Materialia **68**, 150–158 (2014)
2. D. Doraiswamy, S. Ankem, The effect of grain size and stability on ambient temperature tensile and creep deformation in metastable beta titanium alloys. Acta Materialia **51**, 1607–1619 (2003)
3. D. Grevey, V. Vignal, I. Bendaoud, P. Erazmus-Vignal, I. Tomashuchuk, D. Daloz, P. Sallamand, Microstructure and micro-electrochemical study of a tantalum-titanium weld interface. Mater. Des. **87**, 974–985 (2015)
4. C. Kenel, C. Leinenbach, Influence of cooling rate on microstructure formation during rapid solidification of binary TiAl alloys. J. Alloy. Compd. **637**, 242–247 (2015)
5. J.I. Qazi, H.J. Rack, Metastable beta titanium alloys for orthopedic applications. Adv. Eng. Mater. **7**, 993–998 (2005)
6. D.Q. Zhang, Z.H. Liu, Q.Z. Cai, J.H. Liu, C.K. Chua, Influence of Ni content on microstructure of W-Ni alloy produced by selective laser melting. Int. J. Refract. Met. Hard Mater. **45**, 15–22 (2014)
7. Y. Zhu, D. Liu, X. Tian, H. Tang, H. Wang, Characterization of microstructure and mechanical properties of laser melting deposited Ti-6.5Al-3.5Mo-1.5Zr-0.3Si titanium alloy. Mater. Des. **56**, 445–453 (2014)
8. A. Hussien, L. Hao, C. Yan, R. Everson, Finite element simulation of the temperature and stress fields in single layers build without-support in selective laser melting. Mater. Des. **52**, 638–647 (2013)
9. L.E. Loh, C.K. Chua, W.Y. Yeong, J. Song, M. Mapar, S.L. Sing, Z.H. Liu, D.Q. Zhang, Numerical investigation and an effective modelling on the Selective Laser Melting (SLM) process with aluminium alloy 6061. Int. J. Heat Mass Transf. **80**, 288–300 (2015)
10. D. Gu, Y.-C. Hagedorn, W. Meiners, G. Meng, R.J.S. Batista, K. Wissenbach, R. Poprawe, Densification behavior, microstructure evolution, and wear performance of selective laser melting processed commercially pure titanium. Acta Materialia **60**, 3849–3860 (2012)
11. L. Thijs, F. Verhaeghe, T. Craeghs, J. Van Humbeeck, J.P. Kruth, A study of the microstructural evolution during selective laser melting of Ti-6Al-4 V. Acta Materialia **58**, 3303–3312 (2010)
12. Q. Guo, Y. Zhan, H. Mo, G. Zhang, Aging response of the Ti-Nb system biomaterials with β-stabilizing elements. Mater. Des. **31**, 4842–4846 (2010)
13. E. Chlebus, B. Kuznicka, T. Kurzynowski, B. Dybala, Microstructure and mechanical behaviour of Ti-6Al-7Nb alloy produced by selective laser melting. Mater. Charact. **62**, 488–495 (2011)
14. M. Niinomi, Mechanical properties of biomedical titanium alloys. Mater. Sci. Eng. A **243**, 231–236 (1998)

15. H.K. Rafi, N.V. Karthik, H. Gong, T.L. Starr, B.E. Stucker, Microstructures and mechanical properties of Ti6Al4V parts fabricated by selective laser melting and electron beam melting. J. Mater. Eng. Perform. **22**, 3872–3883 (2013)
16. Y.-L. Zhou, M. Niinomi, T. Akahori, M. Nakai, H. Fukui, Comparison of various properties between titanium-tantalum alloy and pure titanium for biomedical applications. Mater. Trans. **48**, 380–384 (2007)
17. Y.L. Zhou, M. Niinomi, T. Akahori, Effects of Ta content on Young's modulus and tensile properties of binary Ti-Ta alloys for biomedical applications. Mater. Sci. Eng. A **371**, 283–290 (2004)

Chapter 5
Statistical Modelling of Selective Laser Melting of Cellular Lattice Structures

The design freedom from selective laser melting (SLM) comes with associated complexity. The modelling of the process is complicated as the physical phenomena occur over a broad range of length and time scales. Three-dimensional (3D) computer simulations, such as to understand the relationship between processing parameters and the thermal behaviour of the material as it is melted by the laser can be quite expensive to run, even on high performance computer systems. This is especially true if they include various aspects of the physics underlying SLM. Exploring the design space using experiments can also be challenging as there are many parameters, as mentioned in previous chapters, which influence the process and thus the final quality of the part. Statistical modelling allows an inexpensive method in analysing the key factors in influencing the parts quality and mechanical properties. The use of design of experiments (DOE) techniques such as the regression analysis and statistical analysis using the analysis of variance (ANOVA), have been shown to be useful approaches to study the effect of many parameters in material processing applications [1].

5.1 Forming Mechanisms of Lattice Structures Using Selective Laser Melting

Various studies have been conducted in the fabrication of cellular lattice structures using SLM. This process has shown great potential in this area. Yan et al. evaluated the manufacturability and performance of SLM produced periodic cellular lattice structures. The lattice structures are designed using gyroid consisting of circular struts and a spherical core as repeating unit. The effect of unit cell size on the manufacturability, density and compression properties of the manufactured structures was investigated [2, 3]. Van Bael et al. evaluated the controllability of porous Ti6Al4V structures fabrication by comparing the fabricated parts to the design. Using pore size, strut thickness, porosity, surface area and structure volume, the SLM robustness is

S. L. Sing, *Selective Laser Melting of Novel Titanium-Tantalum Alloy as Orthopaedic Biomaterial*, Springer Theses, https://doi.org/10.1007/978-981-13-2724-7_5

evaluated [4]. Lin et al. investigated the radiographic characteristics and mechanical properties of a topologically optimized Ti6Al4V lumbar interbody fusion cage made by SLM [5].

SLM is a layer-by-layer additive manufacturing (AM) process, hence, the struts formation in different directions undergo different mechanisms. For the horizontal struts, they are formed by single or multiple continuous melt tracks, depending on the strut dimensions. For the vertical struts, they are formed by direct stacking of single or multiple melt pools across multiple layers. For struts inclined at an angle, for example, diagonal struts, they are formed by stacking of single or multiple melt pools across multiple layers that are offset from each other. The offset depends on the incline angle of the struts.

The various strut forming mechanisms are shown in Fig. 5.1.

Despite the studies conducted, there is limited information on the effect of processing parameters of SLM on the quality and mechanical properties of cellular lattice structures, which this chapter will address.

5.2 Design of Cellular Lattice Structures

The cellular lattice structures used in this chapter are specially designed to study the effect of processing parameters on the quality and mechanical properties of these structures. The unit cell designed consists of vertical, horizontal and diagonal square struts of 0.08 mm sides which corresponds to the spot size of the laser in the SLM 250 HL (SLM Solutions Group AG). Vertical, horizontal and diagonal struts are chosen to investigate the different building direction capabilities of SLM.

The dimensions of the repeating unit cell are 1 mm by 1 mm by 1 mm. The generated CAD diagram is shown in Fig. 5.2. The overall dimensions of the lattice structures are 10 mm by 10 mm by 10.5 mm, allowance is given in the height to allow for erosion of materials from electrical discharge wire cutting of the samples from the substrate plate.

5.3 Design of Experiment for Regression Analysis

Regression analysis is used to determine the value of coefficients of the function that cause the function to best fit a set of observed data [6]. There are mainly two types of regression techniques, namely linear and non-linear regression. This method is employed to develop empirical model for predicting output parameters under a set of controlled experimental factors. Regression analysis optimization process involves three major steps [7]:

1. Performing the statistically designed experiments
2. Estimating the coefficients in a mathematical model
3. Predicting the response and examining the adequacy of the model

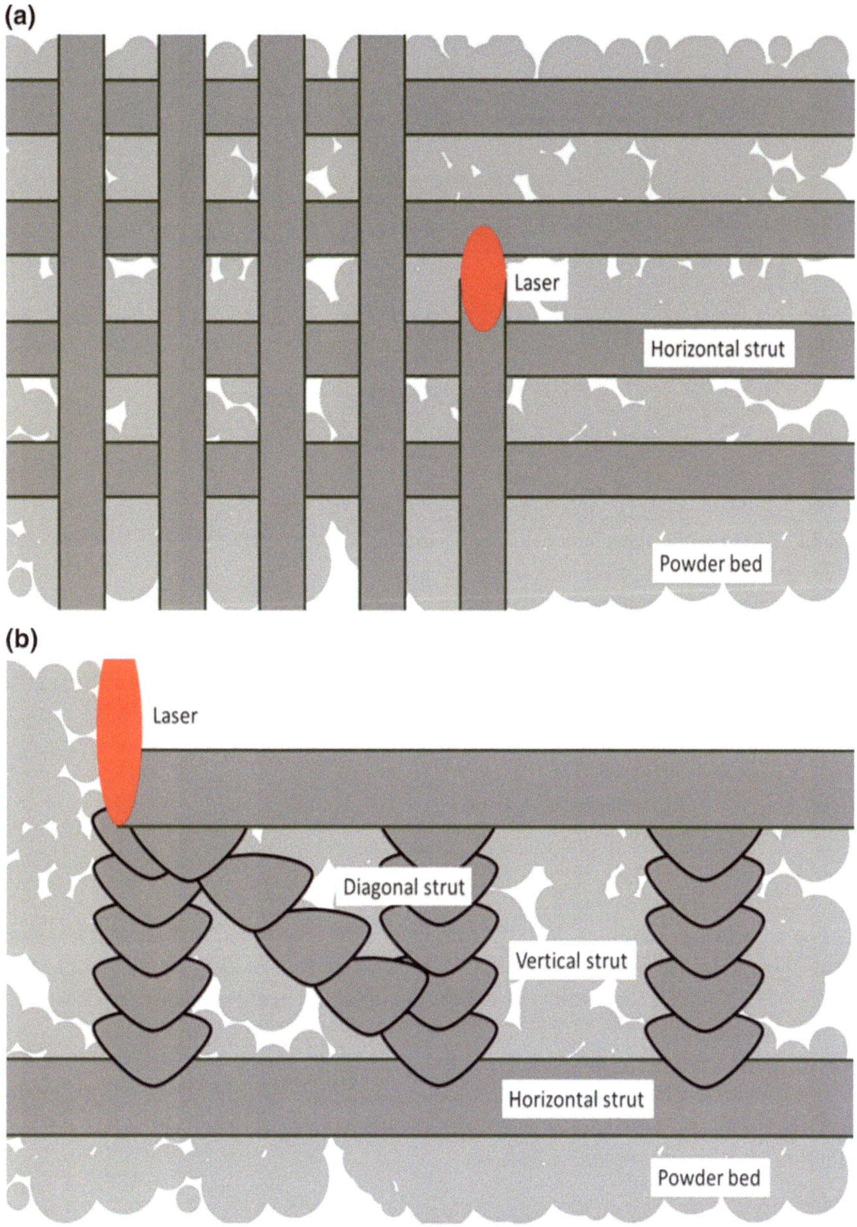

Fig. 5.1 Forming mechanisms for **a** horizontal struts in xy-plane **b** vertical and diagonal struts in yz-plane

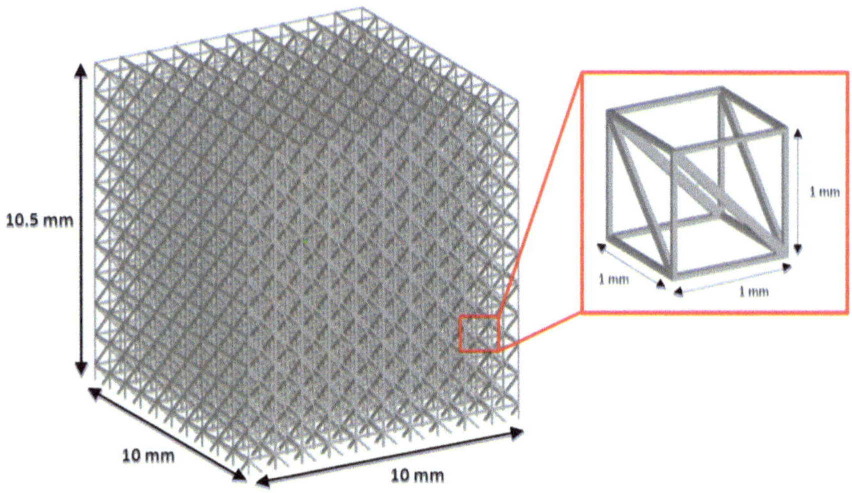

Fig. 5.2 CAD file of cellular lattice structures used for characterisation

Table 5.1 Factors for regression analysis

Factor		Values		
		(−1)	(0)	(1)
P	Laser power (W)	120	240	360
S	Laser scanning speed (mm/s)	400	800	1200
L	Layer thickness (mm)	0.030	0.050	0.100

The significant variables, laser power, laser scanning speed and layer thickness, were chosen as the critical variables designated as P, S and L, respectively. The values of P, S and L were selected within the obtained energy density, as described in previous chapters.

The factors and their three levels for the 3^3 factorial design are listed in Table 5.1.

In order to evaluate the SLM key factors and their effects on the strut dimensions, porosity and mechanical properties of cellular lattice structures, a polynomial equation [1, 8] is expressed as follows:

$$y = a_0 + \sum a_i x_i + \sum a_{ii} x_i^2 + \sum \sum_{i<j} a_{ij} x_i x_j + \varepsilon \tag{5.1}$$

where y is the response or dependent variable investigated (strut dimensions, porosity, yield strength and elastic modulus), a_i is a correction constant coefficient, a_i, a_{ii} and a_{ij} are coefficients for linear, quadratic and interaction effect, x_i and x_j is the independent variables (laser power, scanning speed and layer thickness) and ε is the random error. The polynomial equation assumes that third order interactions of the independent variables are insignificant. ANOVA is carried out on the derived empirical formulae to evaluate the significance of the formulae. ANOVA uses a test statistic called the F statistic to test the null hypotheses. From the F distribution table, with probability of 5%, if the F value of the formula is lower than critical value, the formula is then concluded to be significant in relating the inputs and output. The p value is the probability that the derived empirical formula is not significant where R^2 is a measurement of the fit of the data from 0 to 1 with 1 being the perfect fit.

5.4 Properties Modelling of Cellular Lattice Structures

5.4.1 Metrological Characterisation

The overall dimensions of the as-fabricated lattice structures were measured using digital Vernier callipers with 0.01 mm accuracy (ABS Digimatic Calipers, Mitutoyo Corporation). The sample dimensions were derived from the average of three points (n = 3) on each of the three replicates (N = 3) of the as-fabricated samples. Dry weighing occurred under normal atmosphere conditions using a XS Analytical Balance with sensitivity of 0.001 g and repeatability of 0.0001 g (XS 204, Mettler Toledo), and the density of the samples ρ_{abs} was calculated by dividing the actual weight by the volume of the parts. The porosity of the parts is obtained using the formula as follows:

$$Porosity(\%) = \left(1 - \frac{\rho_{abs}}{\rho_{theoretical}}\right) \times 100 \tag{5.2}$$

where $\rho_{theoretical}$ is the theoretical density of the bulk material.

The struts of the as-fabricated samples underwent morphological characterization using optical microscope (OM, SZX7, Olympus). The same equipment was also used for measurement of the strut dimensions using the OM images. The strut dimensions were measured based on the fully formed strut, without taking into consideration powder adhesion to the struts. For every OM image, 15 values of the strut dimensions were measured (n = 15) and the average value was taken.

The as-fabricated samples replicate the shape of the CAD files designed for this experiment, as shown in Fig. 5.3. The samples show that the designed CAD model can be fabricated successfully using SLM.

The built samples (n = 27, N = 3) have average length of 10.01 ± 0.27 mm, average width of 10.01 ± 0.27 mm and average height of 9.85 ± 0.35 mm. The results showed that the SLM process have high accuracy (shown by the average values which are close to the designed value) and high precision (shown by the small standard deviations). However, the individual samples may deviate slightly from the

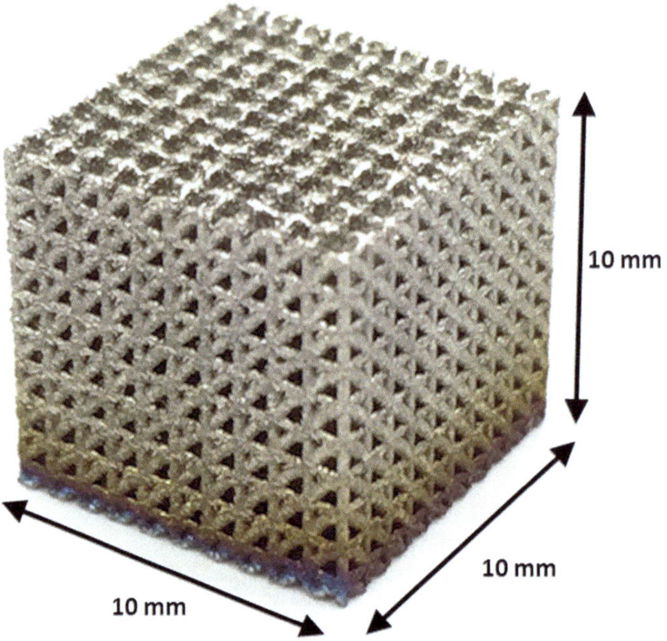

Fig. 5.3 SLM fabricated lattice structure

designed value because the different set of processing parameters used which will results in different melt pool size.

Figure 5.4 shows the OM images of the as-fabricated samples. Sample measurements of strut dimensions were also marked in red in the figure. It is observed that the struts of the lattice structures are solid, connected and continuous, indicating good melting of the powder during the SLM process. The variation between the strut designs in CAD and the actual struts may be due to the laser power fluctuations inherent during the SLM process and the different powder particle sizes.

The actual strut dimensions of the lattice structures for layer thickness 0.03 mm, 0.05 mm and 0.10 mm are tabulated in Fig. 5.5, where H_T is the strut dimension of horizontal strut in xy-plane (perpendicular to build direction, along scanning direction), H_S is the strut dimension of horizontal strut in yz-plane (along build direction, perpendicular to scanning direction), V is the vertical strut dimension and D is the diagonal strut dimension.

A design limit of 0.920 mm, corresponding to the designed pore size, is set as the arbitrary value for struts that are formed across the whole designed pore and 0.000 mm is set as the arbitrary value if the struts are not formed continuously. Examples of these are shown in Fig. 5.6.

The experimental strut dimensions are all found to be larger than the designed values of 0.080 mm. The larger experimental strut dimensions compared to the designed value can be attributed to the following reasons. Firstly, the CAD is sliced and scanned

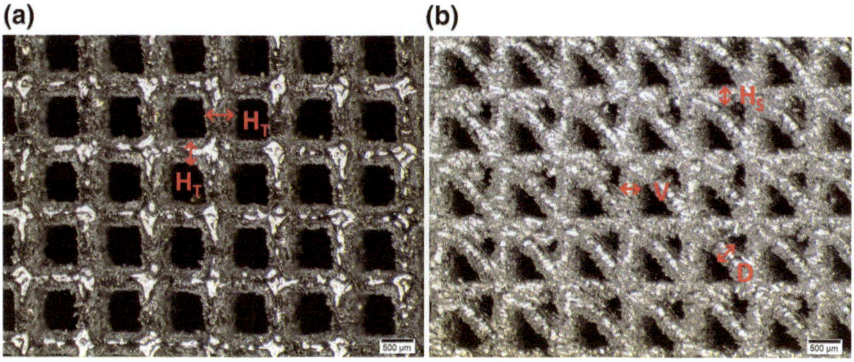

Fig. 5.4 OM images of fully formed struts in lattice structures **a** xy-plane **b** yz-plane

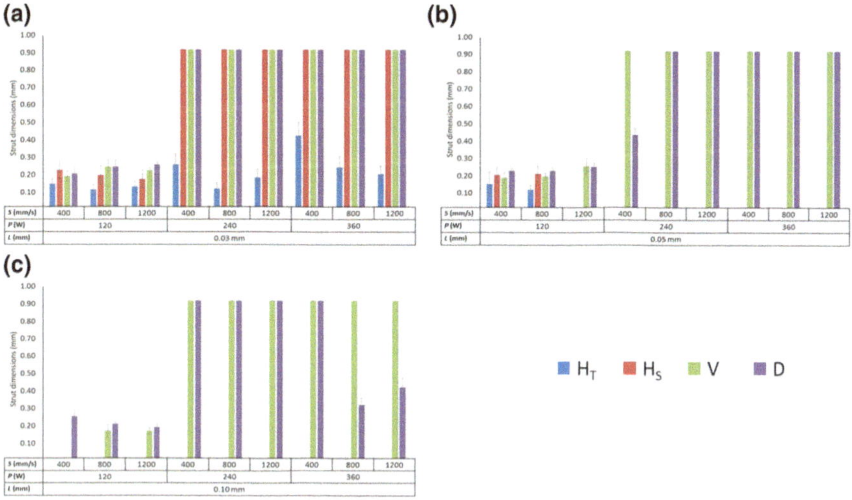

Fig. 5.5 Actual strut dimensions for **a** layer thickness = 0.030 mm **b** layer thickness = 0.050 mm **c** layer thickness = 0.100 mm

using the designed scan strategy which results in strut formations due to a single line scan alone the powder bed. The melt pool formed may have width and depth that are larger than the designed strut dimensions of 0.080 mm.

The forming of the melt pool and melt pool dimensions are shown in Fig. 5.7.

Secondly, the scan results in partially melted metal particles bonded onto the struts, in addition to continuous struts formation. This leads to powder adhesion on the struts. The forming mechanism of powder adhesions onto the struts is illustrated in Fig. 5.8.

The results obtained agree with some of the findings made when porous metal structures were manufactured by SLM. Yan et al. also found the strut sizes of the SLM manufactured gyroid lattice structures were higher than the designed value [2].

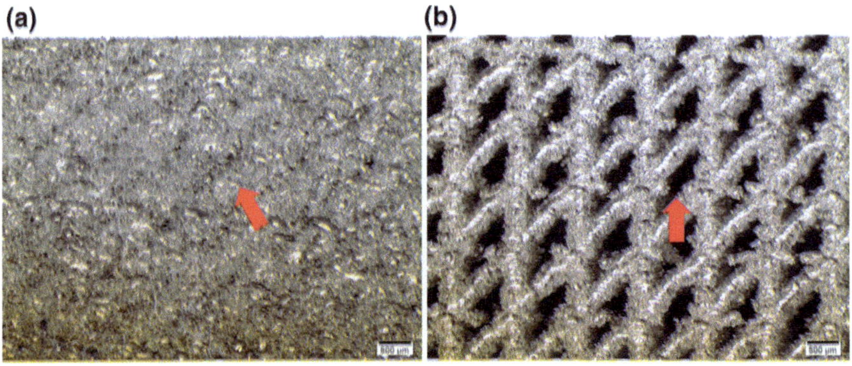

Fig. 5.6 OM images of lattice struts **a** formed across the whole pore **b** not formed continuously

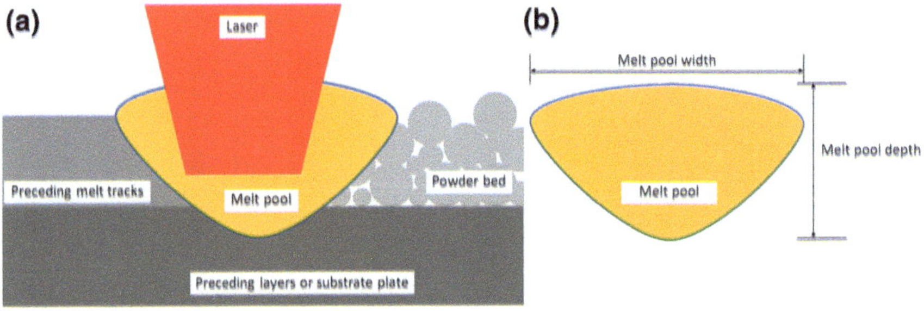

Fig. 5.7 **a** Melt pool formation **b** melt pool dimensions

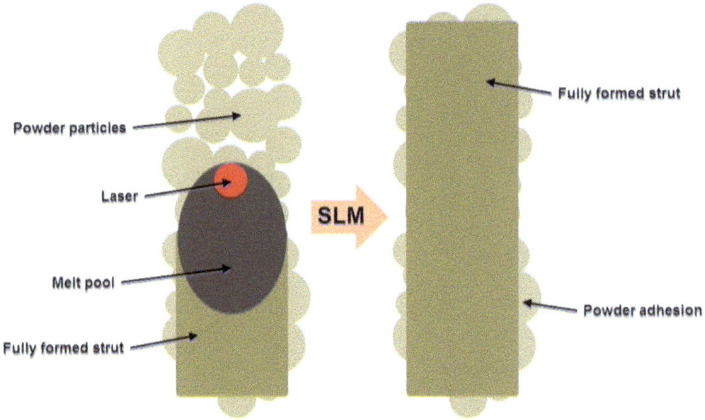

Fig. 5.8 Powder adhesions on struts

Van Bael et al. evaluated the Ti6Al4V porous structure fabricated by SLM through micro-CT image analysis and reported an increase in strut size compared to the designed value as well [4].

From the OM images, it can be observed that there are adhered powder particles on the struts of the lattice structures. Similar phenomenon has also been observed in the titanium porous structures fabricated using SLM by Pattanayak et al. [9]. The presence of these particles on the struts may be due to partial melting phenomenon of the raw titanium particles on the boundary of the solid struts due to heat transfer from the melt pool to the surrounding powder bed. These partially melted particles then adhere to the fully formed struts.

Heat treatment can be carried out to make the partially melted powder particles fuse and bond with the laser melted struts. However, heat treatment may result in texturing and thickening of the struts [9]. Due to powder adhesion on the struts of lattice structures, a different set of parameters may be required for lattice structures fabrication as compared to bulk parts fabrication by SLM which have larger cross sections.

Fluctuations in the strut dimensions, even within a single strut, are observed. This is due to the inherent laser power fluctuation in the SLM machine, even during continuous scan of a single strut. Furthermore, the powder used has a range of sizes which will also result in variation in the strut dimensions since there is only a single scan track per strut.

Statistical Modelling of Results

The second-order regression models were obtained for the strut dimensions H_T, H_S, V and D. Mathematical models developed for this study considered only significant model terms. From the statistical analysis, insignificant model terms which have limited influence were removed to make the empirical formulae efficient [8].

For H_T, the model F value of 8.46 and p-value of 9.72×10^{-5} indicates that the model is significant ($p < 0.05$) [10]. There is only 5% chance that a model F value could become this large due to noise. If the p-value of the factor is larger than 0.05, the factor is not significant in affecting H_T, hence in this case, L is the significant factor. The second order interactions, L^2 and interaction between P and L are significant as well. The predicted R^2 (0.82) indicate a good fit of the experimental data to the developed empirical formula and is in good agreement with the adjusted R^2 (0.72). The empirical formula for H_T in terms of the key SLM parameters and their coefficient is:

$$H_T = -103.54L + 72.83L^2 - 39.72P.L \tag{5.3}$$

For H_S, the model F value of 12.66 and p-value of 6.54×10^{-6} indicates that the model is significant ($p < 0.05$) [10]. There is only 5% chance that a model F value could become this large due to noise. If the p-value of the factor is larger than 0.05, the factor is not significant in affecting H_S, hence in this case, P and L are the significant factors. The second order interactions, L^2 and interaction between P and L are significant as well. The predicted R^2 (0.87) indicates a good fit of the experimental data to the developed empirical formula and is in good agreement with the adjusted R^2 (0.80).The empirical formula for H_S in terms of the key SLM parameters and their coefficient is:

$$H_S = 96.31P - 340.37L + 293.74L^2 - 179.44P.L \quad (5.4)$$

For V, the model F value of 357.73 and p-value of 1.43×10^{-17} indicates that the model is significant ($p < 0.05$) [10]. There is only 5% chance that a model F value could become this large due to noise. If the p-value of the factor is larger than 0.05, the factor is not significant in affecting V, hence in this case, P and L are the significant factors. The second order interactions, P^2, between P and L and between P and S are significant as well. The predicted R^2 (0.99) indicates a good fit of the experimental data to the developed empirical formula and is in good agreement with the adjusted R^2 (0.99). The empirical formula for V in terms of the key SLM parameters and their coefficient is:

$$V = 937.74 + 367.31P - 17.99L - 367.32P^2 + 26.98P.L - 22.83P.S \quad (5.5)$$

For D, the model F value of 9.76 and p-value of 3.85×10^{-5} indicates that the model is significant ($p < 0.05$) [10]. There is only 5% chance that a model F value could become this large due to noise. If the p-value of the factor is larger than 0.05, the factor is not significant in affecting D, hence in this case, only P is the significant factor. The second order interactions, P^2 is significant as well. The predicted R^2 (0.83) indicates a good fit of the experimental data to the developed empirical formula and is in good agreement with the adjusted R^2 (0.75). The empirical formula for D in terms of the key SLM parameters and their coefficient is:

$$D = 863.60 + 282.89P - 350.27P^2 \quad (5.6)$$

Based on the empirical formulae obtained from regression analysis, it is observed that laser power P has the most significant effect on the strut dimensions and it is interesting to note that scanning speed S only affects the strut dimensions on the yz-plane which is along the build direction and perpendicular to the laser scanning direction. Furthermore, layer thickness L has significant effect on both xy-plane which is parallel to the scanning direction, perpendicular to the build direction and yz-plane. The derived formula for the four strut dimensions are all different due to the different forming mechanisms of the struts in different directions and planes as described previously. Such forming mechanisms results in strut dimensions that are highly dependent on the melt pool size, such as the melt pool depth and melt pool width. Despite the 80 μm laser spot size, the melt pool formed will have larger depth and width than the spot size [11–15], which is the designed strut dimensions value. A schematic of the forming mechanisms of these struts in the yz/xz-plane are again shown in Fig. 5.9.

Sensitivity analysis is a method to rank the significant process parameters in order of importance and is used to determine the most influential input process parameters with respect to the outputs. Hence, SLM of lattice structures can be improved by adjusting the setting of these parameters carefully. Mathematically, sensitivity of an output with respect to an input factor is the partial derivative of the output with

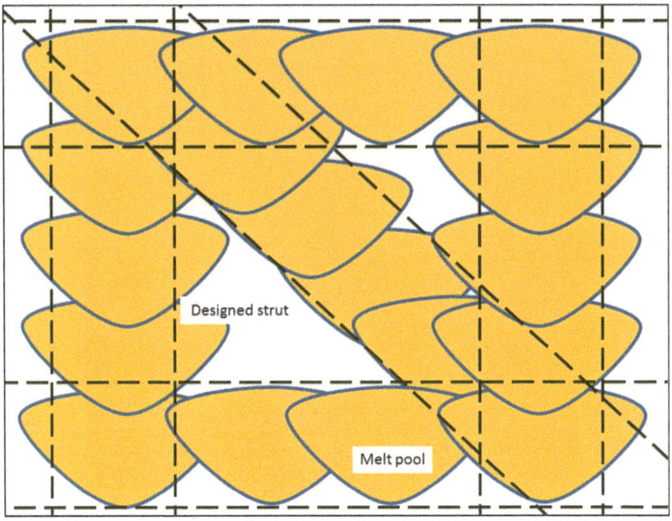

Fig. 5.9 Forming mechanisms of struts in yz/xz-plane

respect to the input variables [16, 17]. The larger the magnitude of the sensitive of an output with respect to an input factor, the more significant the effect of the change in input on the output. If the sensitivity with respect to a certain process parameter is positive, the strut dimensions will increase with an increase in this process parameter, whereas negative sensitivity states the opposite [8].

The sensitivity analysis results are shown in Figs. 5.10 and 5.11.

It is revealed that the strut dimensions are most sensitive to layer thickness, as compared to laser power and scanning speed for the horizontal strut. In contrast, the strut dimensions of the vertical struts and diagonal struts are most sensitive to laser power, as compared to layer thickness and scanning speed. These are shown by the magnitude of the sensitivity in the analysis.

Furthermore, it is also shown that D is only sensitive to laser power and V is highly sensitive to laser power as compared to layer thickness and scanning speed. The results imply that for better control of the vertical and diagonal struts dimensions, there is a need for better control of the laser power, as compared to varying layer thickness and scanning speed.

5.4.2 Mechanical Characterisation

The fabricated cubic samples have designed dimensions of 10 mm by 10 mm by 10 mm, which are used as test coupons for compression tests based on ISO 13314:2011 (Mechanical testing of metals—Ductility testing—Compression test for porous and cellular metals).

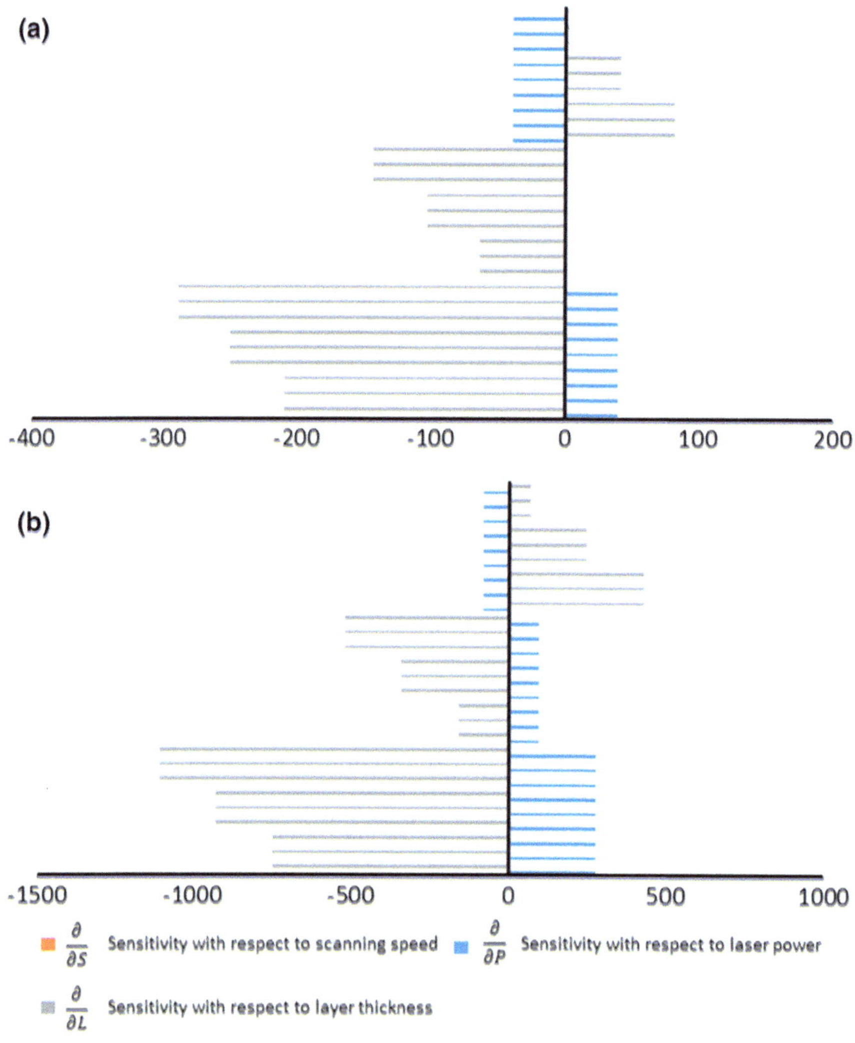

Fig. 5.10 Sensitivity of strut dimensions to SLM process parameters **a** H_T **b** H_S

Uni-axial compression tests were carried out, at room temperature (25 °C), to assess the compressive properties of the lattice structures, each with three replicates (n = 3), by using Instron Static Tester Series 5569 equipped with a 50 kN load cell. The loading speed was set at a constant of 0.6 mm/min, to maintain a constant strain rate for all tests as recommended by the standard. The compressive deformation rate has to be set such that the strain rate experienced by the samples are constant throughout.

The compressive deformation rate and strain rate is related by the following equation:

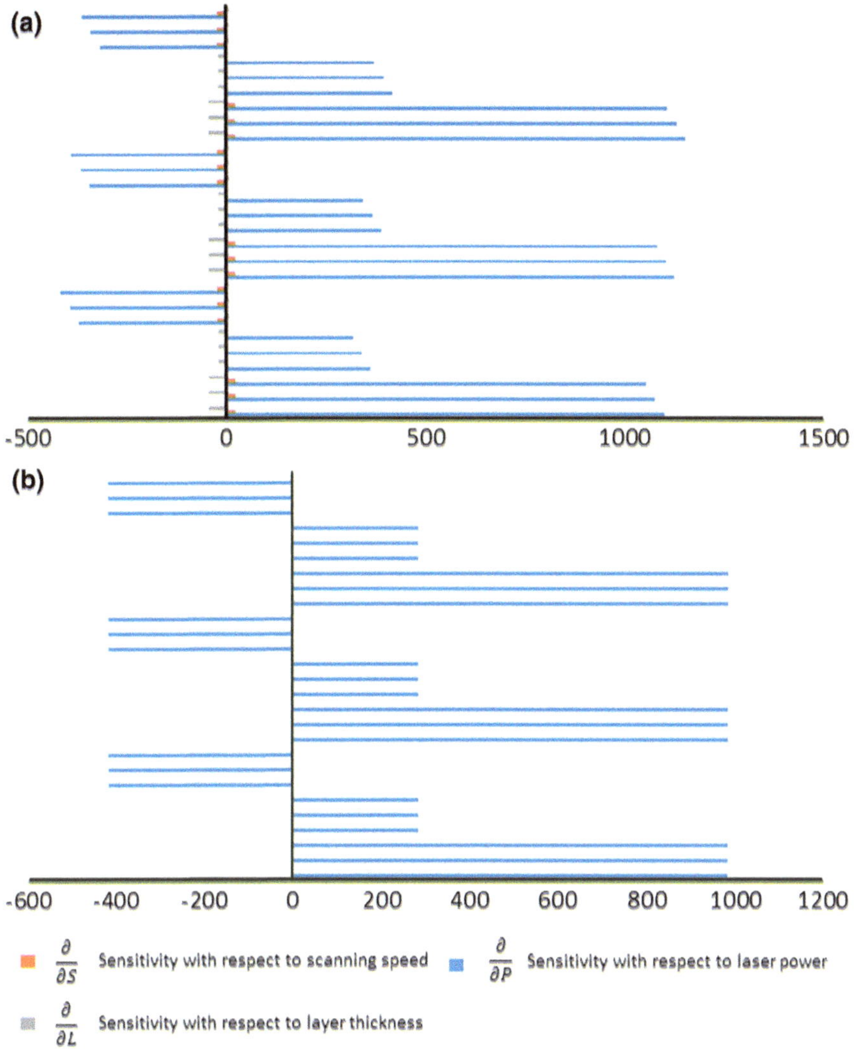

Fig. 5.11 Sensitivity of strut dimensions to SLM process parameters **a** V **b** D

$$strain\,rate = \frac{compressive\,deformation\,rate}{height\,of\,test\,sample} \tag{5.7}$$

Using a standardized strain rate minimizes the differences in results obtained due to the effect of strain rate on hardening behaviour which may be present in some materials [18–21]. ISO 13314:2011 recommends a range of strain rates between 10^{-3} and 10^{-2}/s (0.06 and 0.6/min). To ensure consistency, the test samples are all strained at a uniform rate, without sudden changes, to minimize the effects of different strain rates [19, 20, 22]. The low strain rate of 10^{-3}/s is used as a high

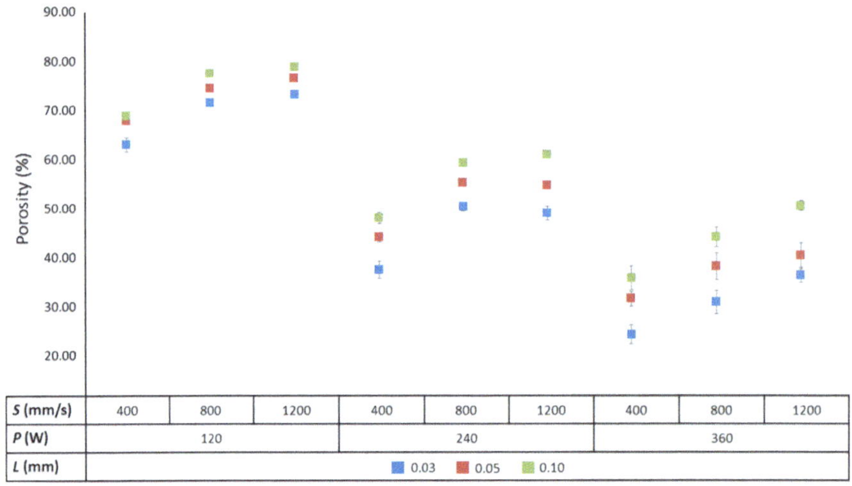

Fig. 5.12 Porosity of lattice structures

strain rate recommended by ISO 13314:2011 can lead to abnormal strain hardening behaviour in titanium [18–20].

The compression tests were carried out until axial deformation of the samples was equal to 100% or when the maximum loading of 50 kN was reached, whichever came first. The stress-strain curves, yield strengths and elastic constants in compression of the as-fabricated samples were then obtained through the compression tests.

The resulting porosity, elastic constant in compression and yield strength of the as-fabricated lattice structures are shown in Fig. 5.12, Fig. 5.13 and Fig. 5.14 respectively. The slope of the straight-line portion of the stress-strain curve is established to define the elastic constant and the yield strength is taken as the stress at plastic compressive strain of 0.2%. The standard deviation in the elastic constant and yield strength may be due to the varying amount of powder adhesion on the struts, which consequently affects the compressive properties of the lattice structures.

Due to the effect of the SLM process parameters, the elastic constant of the lattice structures (n = 27, N = 3) can range from 1.36 ± 0.11 to 6.82 ± 0.15 GPa. This shows that there is a need for careful control of the process parameters during the fabrication of lattice structures to obtain the desired mechanical properties. It also shows the versatility of TiTa lattice structures in orthopaedic applications where bones have wide range of elastic constants [23, 24].

Using the Gibson–Ashby model [25], the elastic modulus and yield strength at different porosities can be estimated as follows:

$$\frac{E}{E_0} = C_E\left(\frac{\rho}{\rho_0}\right)^{k_E} = C_E(1 - P_r)^{k_E} \tag{5.8}$$

$$\frac{Y_s}{Y_0} = C_Y\left(\frac{\rho}{\rho_0}\right)^{k_Y} = C_Y(1 - P_r)^{k_Y} \tag{5.9}$$

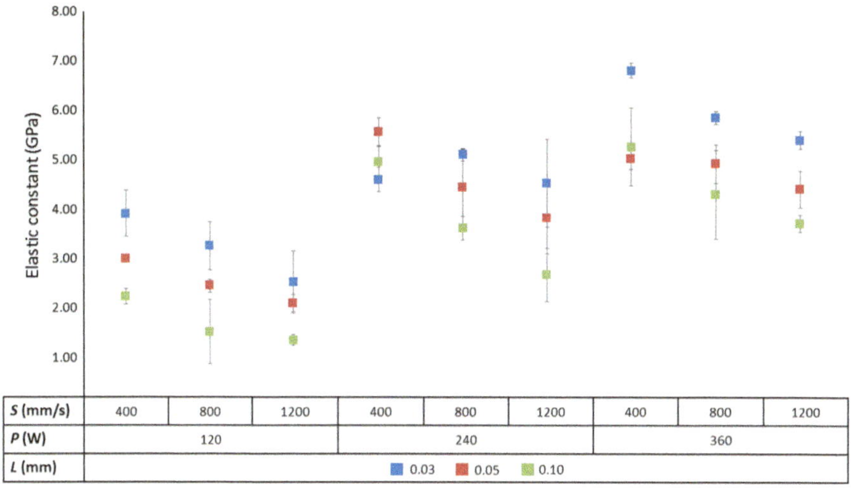

Fig. 5.13 Elastic constant of lattice structures

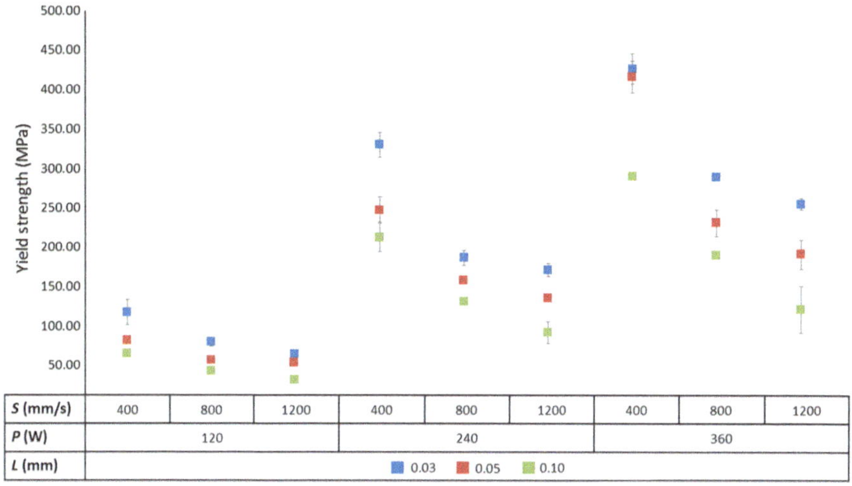

Fig. 5.14 Yield strength of lattice structures

where E, Y_s and ρ are the theoretical elastic constant, yield strength and density of the lattice structures, E_0, Y_0 and ρ_0 are the elastic constant, yield strength and density of fully dense material respectively. P_r is the porosity of the lattice structures. C_E, C_Y, k_E and k_Y are constants that can be calculated based on the compression test results after fitting the formulae. The elastic constant and yield strength is taken to be 75.77 ± 4.04 GPa and 882.77 ± 19.60 MPa respectively from the fully dense TiTa (n = 5). The value of the constants is calculated to be $C_E = 0.112$, $k_E = 0.996$, $C_Y = 0.699$ and $k_Y = 1.785$.

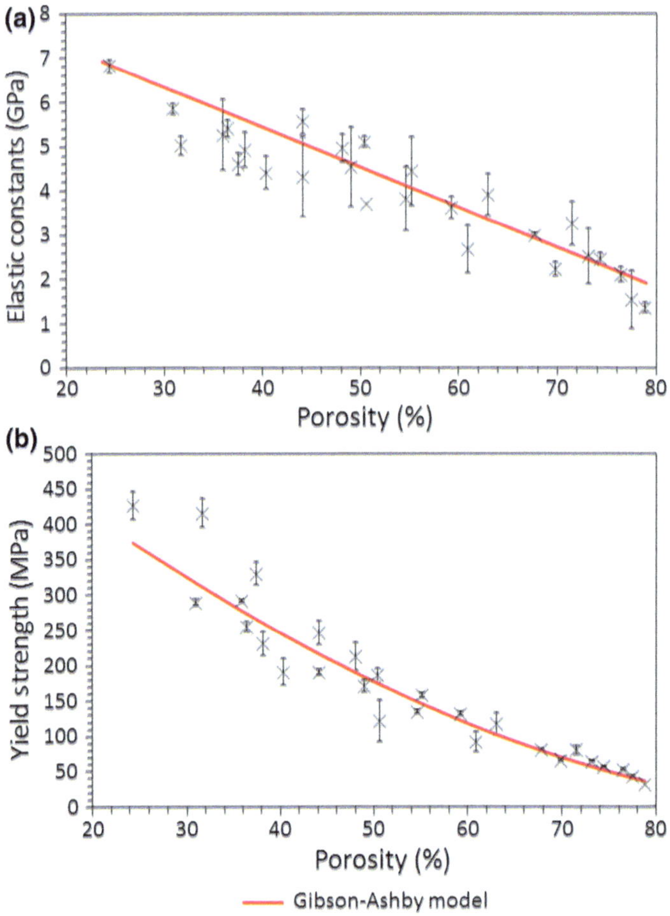

Fig. 5.15 Theoretical and experimental values of **a** elastic constant **b** yield strength of lattice structures

The experimental values of elastic constant, yield strength and their corresponding theoretical values based on the Gibson-Ashby model is plotted in Fig. 5.15.

It is observed that the elastic modulus and yield strength of the as-fabricated lattice structures decrease with increase in porosity, which is consistent with the prediction from the Gibson-Ashby model. From Fig. 5.15, there are differences between experimentally tested and Gibson-Ashby model estimated elastic modulus and yield strength. The differences between the theoretical and experimental values may be attributed to the residual stress inherent due the SLM process, waviness and roughness of the strut surfaces [3]. It can also be due to the disregard of the SLM process parameters which affects the powder adhesions onto the struts which will affect the compressive properties of the lattice structures. The model also failed to consider the type of failure underwent by the lattice structures. This may be due to limitation

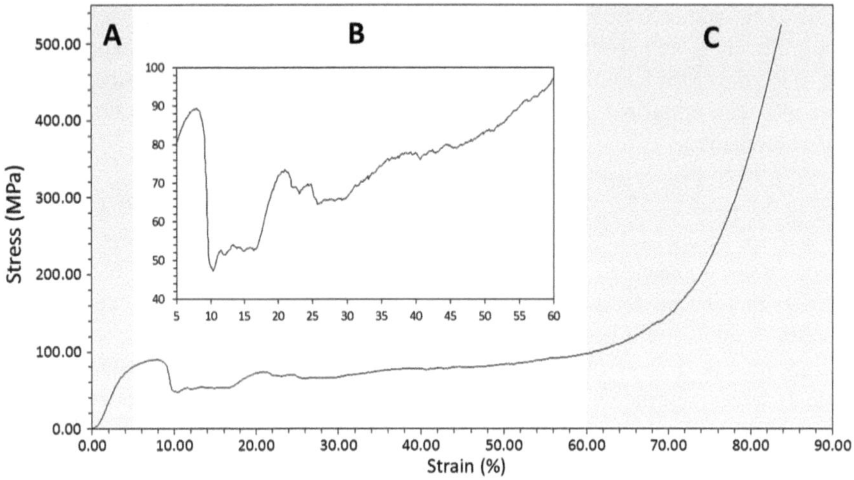

Fig. 5.16 Typical stress-strain curve of lattice structures

of the model, which is designed for porous structures fabricated using conventional methods. Furthermore, the Gibson-Ashby model parameter C in the equation have subsumed all the geometrical information, which includes unit cell type, size of the repeating unit cell and strut dimensions, for the lattice structures except the porosity. Hence, these lacks of geometrical information accounted in the model can results in deviation between the calculated theoretical value and experimental data.

A typical stress-strain curve of the lattice structures is shown in Fig. 5.16.

It can be observed that the SLM fabricated lattice structures made from the designed unit cells have an elastic deformation region (region A), followed by a plateau region in which the stress fluctuates (region B, shown in insert of Fig. 5.16), and finally a densification region characterized by a rapid stress increase (region C). Buckling occurs from region B and results in stress fluctuation in this region which is a characteristic of brittle fracture. The samples underwent permanent deformation and fracture after the compression. The brittle behaviour is also observed in Ti6Al4V cellular structures fabricated by electron beam melting (EBM) [26].

Statistical Modelling of Results

The second-order regression models were obtained for the porosity, elastic constant and yield strength of the lattice structures. Mathematical models developed for this study considered only significant model terms. From the statistical analysis, insignificant model terms which have limited influence were removed to make the empirical formulae efficient [8].

For porosity, the model F value of 330.09 and p-value of 2.81×10^{-17} indicates that the model is significant ($p < 0.05$) [10]. There is only 5% chance that a model F value could become this large due to noise. If the p-value of the factor is larger than 0.05, the factor is not significant in affecting the porosity, hence in this case, P, S and L

are all significant factors. The second order interactions, P^2, S^2 and between P and L are significant as well. The predicted R^2 (0.99) indicate a good fit of the experimental data to the developed empirical formula and is in good agreement with the adjusted R^2 (0.99). The empirical formula for P_r in terms of the key SLM parameters and their coefficient is:

$$\mathrm{P}_r = 53.40 - 17.77P - 5.46S + 4.93L + 3.71P^2 - 3.35S^2 - 1.70P.L \quad (5.10)$$

For elastic constant E, the model F value of 32.10 and p-value of 6.20×10^{-9} indicates that the model is significant ($p < 0.05$) [10]. There is only 5% chance that a model F value could become this large due to noise. If the p-value of the factor is larger than 0.05, the factor is not significant in affecting E, hence in this case, P, S and L are all significant factors. The second order interactions, P^2 is significant as well. The predicted R^2 (0.94) indicates a good fit of the experimental data to the developed empirical formula and is in good agreement with the adjusted R^2 (0.91). The empirical formula for E in terms of the key SLM parameters and their coefficient is:

$$E = 4.33 + 1.30P - 0.60S - 0.69L + 0.59P^2 \quad (5.11)$$

For yield strength Y_s, the model F value of 100.28 and p-value of 6.04×10^{-13} indicates that the model is significant ($p < 0.05$) [10]. There is only 5% chance that a model F value could become this large due to noise. If the p-value of the factor is larger than 0.05, the factor is not significant in affecting Y_s, hence in this case, P, S, and L are all significant factors. The second order interactions, P^2, S^2, and the interaction between P and L, P and S are significant as well. The predicted R^2 (0.98) indicates a good fit of the experimental data to the developed empirical formula and is in good agreement with the adjusted R^2 (0.97). The empirical formula for Y_s in terms of the key SLM parameters and their coefficient is:

$$\begin{aligned} Y_s = {} & 165.51 + 101.09P - 59.60S - 41.30L - 17.94P^2 \\ & + 31.74S^2 - 20.47P.L - 37.45P.S \end{aligned} \quad (5.12)$$

Based on the empirical formulae obtained from regression analysis, it is observed that all three key factors, laser power P, scanning speed S and layer thickness L have significant effect on the mechanical properties of the lattice structures.

Sensitivity analysis is a method to rank the significant process parameters in order of importance and is used in determining the most influential input process parameters with respect to the outputs. Hence, the lattice structures mechanical properties can be controlled by adjusting the setting of these SLM parameters carefully. Mathematically, sensitivity of an output with respect to an input factor is the partial derivative of the output with respect to the input variables [16, 17]. The larger the magnitude of the sensitive of an output with respect to an input factor, the more significant the effect of the change in input on the output. If the sensitivity with respect to a

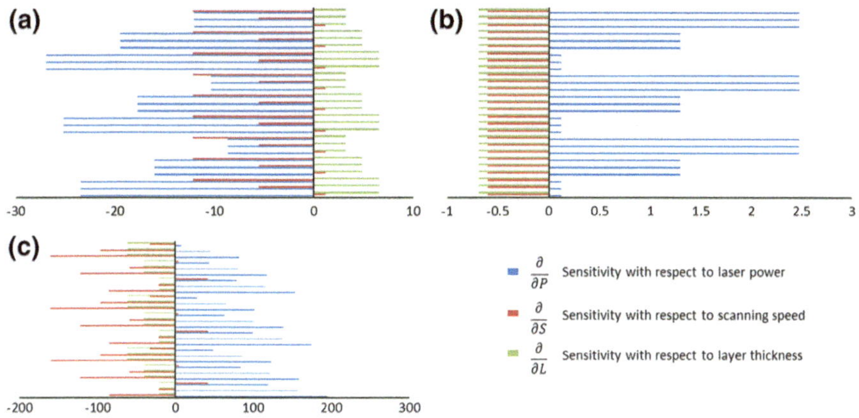

Fig. 5.17 Sensitivity of mechanical properties to SLM process parameters **a** porosity **b** elastic constant **c** yield strength

certain process parameter is positive, the mechanical properties will increase with an increase in this process parameter, whereas negative sensitivity states the opposite [8].

The sensitivity analysis results are shown in Fig. 5.17.

It is revealed that the porosity and elastic constant are most sensitive to laser power, as compared to layer thickness and scanning speed. However, an increase in laser power will lead to a decrease in porosity and an increase in elastic constant. This is in agreement with the findings from the Gibson-Ashby model which states that the porosity and elastic constant has an inverse relationship. For yield strength, it is both sensitive to laser power and scanning speed to almost the same extent, with it being least sensitive to layer thickness.

5.5 Summary

This chapter provides a better understanding of the SLM process parameters that have significant effects on the fabrication of lattice structures. SLM process parameters have been found to have a significant effect on the dimensional accuracy and mechanical properties of the fabricated lattice structures.

Based on the statistical modelling, the following key findings can be summarised:

(1) The regression analysis method can be used to analyse the effect of SLM process parameters on the strut dimensions and mechanical properties of the lattice structures fabricated quantitatively.
(2) By careful manipulation of the process parameters, dimensional accuracy of the lattice structures can be improved. It can also lead to better control of the resulting mechanical properties.

(3) The experimental strut dimensions are found to be larger than the designed value, as shown in Fig. 5.5. This can be attributed to the lower limit of melt pool size during SLM.
(4) The horizontal strut dimensions are most sensitive to layer thickness. In contrast, the vertical and diagonal strut dimensions are most sensitive to laser power. This is shown in Figs. 5.10 and 5.11.
(5) The mechanical properties of SLM fabricated lattice structures are all dependent on laser power, scanning speed and layer thickness. However, the porosity and elastic constant are most sensitive to laser power, as compared to layer thickness and scanning speed. This is shown in Fig. 5.17.

References

1. N. Read, W. Wang, K. Essa, M.M. Attallah, Selective laser melting of AlSi10 Mg alloy: Process optimisation and mechanical properties development. Mater. Des. **65**, 417–424 (2015)
2. C. Yan, L. Hao, A. Hussein, D. Raymont, Evaluations of cellular lattice structures manufactured using selective laser melting. Int. J. Mach. Tools Manuf. **62**, 32–38 (2012)
3. C. Yan, L. Hao, A. Hussein, P. Young, D. Raymont, Advanced lightweight 316L stainless steel cellular lattice structures fabricated via selective laser melting. Mater. Des. **55**, 533–541 (2014)
4. S. Van Bael, G. Kerckhofs, M. Moesen, G. Pyka, J. Schrooten, J.P. Kruth, Micro-CT-based improvement of geometrical and mechanical controllability of selective laser melted Ti6Al4V porous structures. Mater. Sci. Eng. A **528**, 7423–7431 (2011)
5. C.Y. Lin, T. Wirtz, F. LaMarca, S.J. Hollister, Structural and mechanical evaluations of a topology optimized titanium interbody fusion cage fabricated by selective laser melting process. J. Biomed. Mater. Res. Part A **83**, 272–279 (2007)
6. A. Garg, K. Tai, M.M. Savalani, State-of-the-art in empirical modelling of rapid prototyping processes. Rapid Prototyp. J. **20**, 164–178 (2014)
7. C.P. Paul, S.K. Mishra, C.H. Premsingh, P. Bhargava, P. Tiwari, L.M. Kukreja, Studies on laser rapid manufacturing of cross-thin-walled porous structures of Inconel 625. Int. J. Adv. Manuf. Technol. **61**, 757–770 (2012)
8. S. Negi, S. Dhiman, R.K. Sharma, Determining the effect of sintering conditions on mechanical properties of laser sintered glass filled polyamide parts using RSM. Measurement **68**, 205–218 (2015)
9. D.K. Pattanayak, A. Fukuda, T. Matsushita, M. Takemoto, S. Fujibayashi, K. Sasaki, N. Nishida, T. Nakamura, T. Kokubo, Bioactive Ti metal analogous to human cancellous bone: Fabrication by selective laser melting and chemical treatments. Acta Biomaterialia **7**, 1398–1406 (2011)
10. D.C. Montgomery, *Design and Analysis of Experiments*, 7th edn. (Wiley, New York, 2009)
11. L.E. Loh, C.K. Chua, W.Y. Yeong, J. Song, M. Mapar, S.L. Sing, Z.H. Liu, D.Q. Zhang, Numerical investigation and an effective modelling on the Selective Laser Melting (SLM) process with aluminium alloy 6061. Int. J. Heat Mass Transf. **80**, 288–300 (2015)
12. S. Tsopanos, R.A.W. Mines, S. Mckown, Y. Shen, W.J. Cantwell, C.J. Sutcliffe, The influence of processing parameters on the mechanical properties of selectively laser melted stainless steel microlattice structures. J. Manuf. Sci. Eng. **132**, 041011–041011–041012 (2010)
13. I. Yadroitsev, P. Bertrand, I. Smurov, Parametric analysis of selective laser melting process. Appl. Surf. Sci. **253**, 8064–8069 (2007)
14. L. Yadroitsev, I. Shishkovsky, P. Bertrand, I. Smurov, Manufacutring of fine-structured 3D porous filter elements by selective laser melting. Appl. Surf. Sci. **255**, 5523–5527 (2009)
15. L. Yadroitsev, L. Thivillon, P. Bertrand, I. Smurov, Strategy of manufacturing components with designed internal structure by selective laser melting of metallic powder. Appl. Surf. Sci. **254**, 980–983 (2007)

16. I.S. Kim, K.J. Son, Y.S. Yang, P.K.D.V. Yaragada, Sensitivity analysis for process parameters in GMA welding processes using a factorial design method. Int. J. Mach. Tools Manuf. **43**, 763–769 (2003)
17. S. Karaoğlua, A. Seçgin, Sensitivity analysis of submerged arc welding process parameters. J. Mater. Process. Technol. **202**, 500–507 (2008)
18. Y.M. Wang, J.Y. Huang, T. Jiao, Y.T. Zhu, A.V. Hamza, Abnormal strain hardening in nanostructured titanium at high strain rates and large strains. J. Mater. Sci. **42**, 1751–1756 (2007)
19. D.R. Chichili, K.T. Ramesh, K.J. Hemker, The high strain-rate response of alpha-titanium: experiments, deformation mechanisms and modeling. Acta Mater. **46**, 1025–1043 (1998)
20. N.P. Gurao, R. Kapoor, S. Suwas, Deformation behavior of commercially pure titanium at extreme strain rates. Acta Materialia **59**, 3431–3446 (2011)
21. J. Zhang, H. Di, Y. Deng, R.D.K. Misra, Effect of martensite morphology and volume fraction on strain hardening and fracture behavior of martensite-ferrite dual phase steel. Mater. Sci. Eng. A **627**, 230–240 (2015)
22. S. Nemat-Nasser, W.G. Guo, J.Y. Cheng, Mechanical properties and deformation mechanisms of a commercially pure titanium. Acta Materialia **47**, 3705–3720 (1999)
23. P.K.Z. Zysset, X.E. Guo, C.E. Hoffler, K.E. Moore, S.A. Goldstein, Elastic modulus and hardness of cortical and trabecular bone lamellae measured by nanoindentation in the human femur. J. Biomech. **32**, 1005–1012 (1999)
24. B. Piotrowski, A.A. Baptista, E. Patoor, P. Bravetti, A. Eberhardt, P. Laheurte, Interaction of bone-dental implant with new ultra low modulus alloy using a numerical approach. Mater. Sci. Eng.: C **38**, 151–160 (2014)
25. L.J. Gibson, M.F. Ashby, *Cellular Solids: Structures and Properties*, 2nd edn. (Cambridge University Press, New York, 1997)
26. X.Y. Cheng, S.J. Li, L.E. Murr, Z.B. Zhang, Y.L. Hao, R. Yang, F. Medina, R.B. Wicker, Compression deformation behavior of Ti-6Al-4 V alloy with cellular structures fabricated by electron beam melting. J. Mech. Behav. Biomed. Mater. **16**, 153–162 (2012)

Chapter 6
Characterisation of Titanium-Tantalum Lattice Structures Fabricated by Selective Laser Melting

As selective laser melting (SLM) produced titanium-tantalum (TiTa) is new, little is known about the biological response of this material. In this chapter, porous TiTa lattice structures fabricated using SLM are characterised mechanically and biologically. The dimension accuracy of the structures is also investigated using their strut dimensions. The experimental strut dimensions are compared to the computer design aided (CAD) model values. All results are benchmarked against the more established commercially pure titanium (cpTi) and Ti6Al4V produced using the same method.

6.1 Design of Cellular Lattice Structures

The cellular lattice structures used in this chapter is specially designed based on literature review [1–4]. The unit cell is a cross with square strut of 0.285 mm. The dimensions of the repeating unit cell were 1 mm by 1 mm by 1 mm as shown in Fig. 6.1. The SLM capability in building horizontal and vertical struts has been demonstrated in the previous chapter. The materials used in this study are the in-house developed TiTa, cpTi and Ti6Al4V.

6.2 Metrological Characterisation

The overall dimensions of the as-fabricated lattice structures were measured using digital Vernier callipers with 0.01 mm accuracy (ABS Digimatic Calipers, Mitutoyo Corporation). The sample dimensions were derived from the average of three points (n = 3) on each of the three replicates (N = 3) of the as-fabricated samples. Dry weighing occurred under normal atmosphere conditions using a XS Analytical Balance with sensitivity of 0.001 g and repeatability of 0.0001 g (XS 204, Mettler

S. L. Sing, *Selective Laser Melting of Novel Titanium-Tantalum Alloy as Orthopaedic Biomaterial*, Springer Theses, https://doi.org/10.1007/978-981-13-2724-7_6

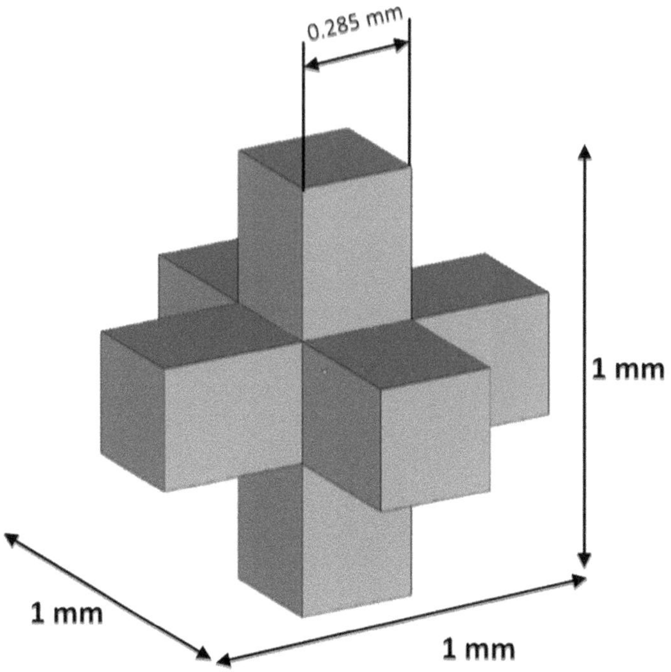

Fig. 6.1 Repeating unit cell used in CAD

Toledo), and the density of the samples ρ_{abs} was calculated by dividing the actual weight by the volume of the parts. The porosity of the parts is obtained using the formula as follows:

$$Porosity(\%) = \left(1 - \frac{\rho_{abs}}{\rho_{theoretical}}\right) \times 100 \quad (6.1)$$

where $\rho_{theoretical}$ is the theoretical density of the bulk material.

The struts of the as-fabricated samples underwent morphological characterization using optical microscope (OM, SZX 7, Olympus). The same equipment was also used for measurement of the strut dimensions using the OM images. The strut dimensions were measured based on the fully formed strut, without taking into consideration powder adhesion to the struts. For every OM image, 15 values of the strut dimensions were measured (n = 15) and the average value was taken.

The generated CAD model is shown in Fig. 6.2. The overall dimension of the lattice structures is 10 mm by 10 mm by 10.5 mm, allowance is given in the height to allow for erosion of materials from electrical discharge wire cutting of the samples from the substrate plate.

The porosity of the samples is obtained using Eq. 6.1. In this study, $\rho_{theoretical}$ is taken to be 7.10, 4.43 and 4.51 g/cm^3 for TiTa, Ti6Al4V and cpTi respectively.

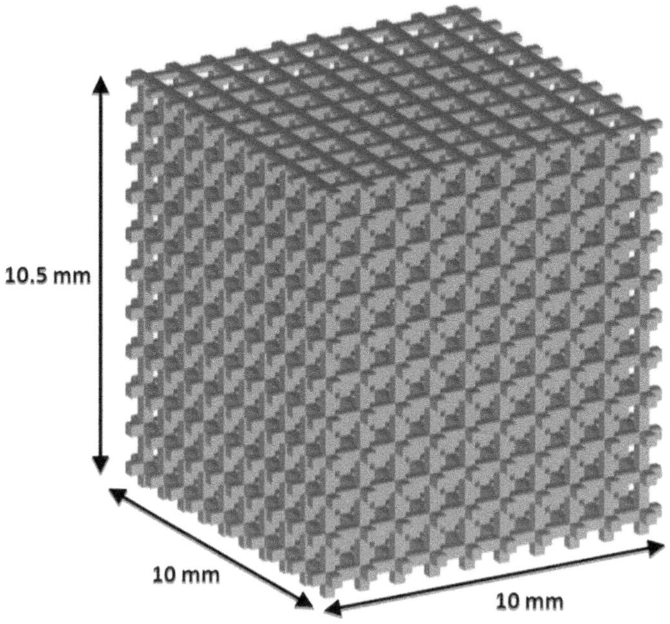

Fig. 6.2 CAD file of cellular lattice structures used for characterisation

The as-fabricated samples replicate the shapes of the CAD files designed for this experiment, as shown in Fig. 6.3. The samples show that the designed CAD model can be fabricated successfully using SLM for TiTa, cpTi and Ti6Al4V.

Sample measurements of strut dimensions are marked in red in Fig. 6.4. It is observed that the struts of the lattice structures are solid, connected and continuous, indicating good powder melting during the SLM process. The variation between the strut designs in CAD and the actual struts may be due to the laser power fluctuations during the SLM process.

The strut dimensions of the lattice structures are tabulated in Table 6.1 and shown in Fig. 6.5 where H_T is the strut dimension of horizontal strut in xy-plane (perpendicular to build direction, along scanning direction), H_S is the strut dimension of horizontal strut in yz-plane (along build direction, perpendicular to scanning direction) and V is the vertical strut dimension.

The experimental strut diameters are found to be larger than the designed values of 0.285 mm. The larger experimental strut diameters compared to the designed value can be attributed to the following reasons. Firstly, the CAD is sliced and scanned using the designed scan strategy with a contour scanning. The contour scan results in partially melted metal particles bonded onto the struts, instead of forming the continuous struts. Secondly, the melt pool size of the laser scan exceeded the boundaries of a strut as the laser spot size of 80 μm is smaller than the designed values which results in melt pool exceeding the designed strut boundaries. This finding agrees with the findings discussed in previous chapters.

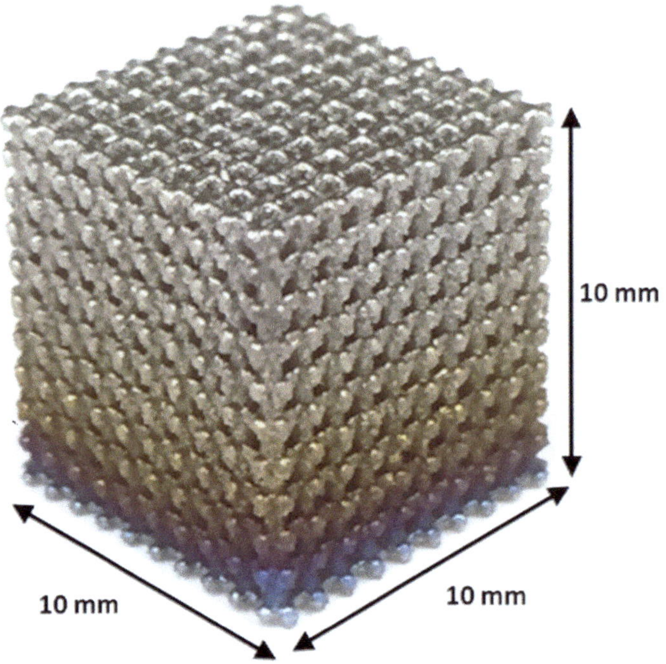

Fig. 6.3 SLM fabricated lattice structure

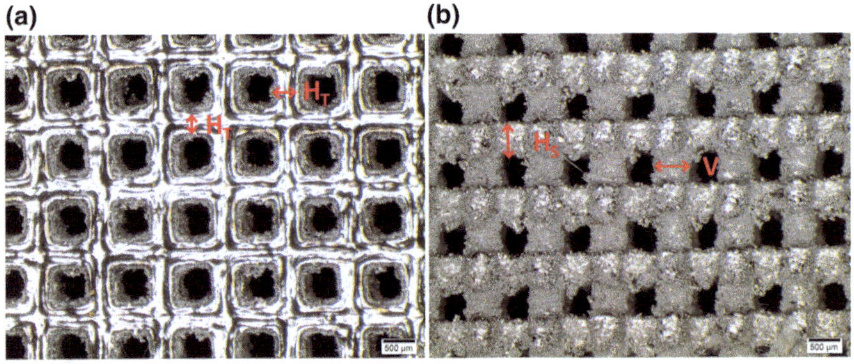

Fig. 6.4 OM images of fully formed struts in lattice structures **a** xy-plane **b** yz-plane

From Fig. 6.5 and ANOVA, it can be seen that there is no significant difference in the strut dimensions H_T and H_S, between the three materials as the SLM process parameters used in fabrication is constant throughout. However, there is a significant difference between V for the three materials. Using Tukey-Kramer test, V for cpTi and Ti6Al4V, Ti6Al4V and TiTa are found to have significant difference. There is no significant difference between V for cpTi and TiTa. This can be due to the lack

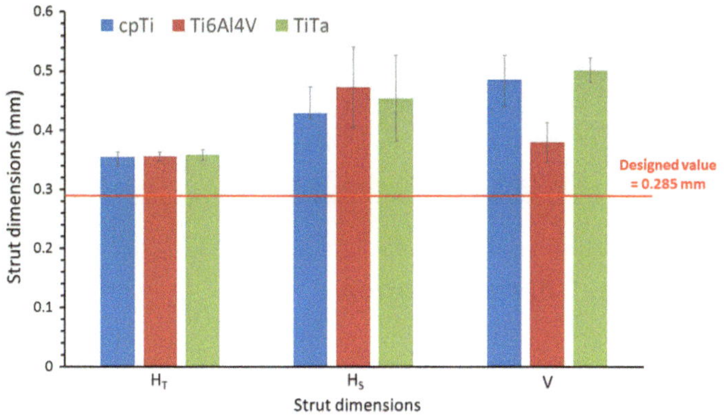

Fig. 6.5 Strut dimensions of TiTa, cpTi and Ti6Al4V lattice structures (n = 10)

Table 6.1 Strut dimensions of TiTa, cpTi and Ti6Al4V lattice structures

Materials	Strut dimensions (mm), n = 10		
	H_T	H_S	V
Designed value	0.285		
TiTa	0.359 ± 0.09	0.454 ± 0.073	0.502 ± 0.021
cpTi	0.356 ± 0.08	0.429 ± 0.045	0.486 ± 0.042
Ti6Al4V	0.357 ± 0.08	0.473 ± 0.069	0.389 ± 0.034

of support structures and melted materials around the struts which lead to lower thermal conductivity due to neighbouring powder. This leads to lower solidification rates as heat is conducted away slower from the melt pool. Lower solidification rate increases instability of the melt pools. As the struts used to measure V are formed by the stacking of melt pools, in contrast to the other strut dimensions, it will be more dependent on the surrounding thermal conductivity. As a result, the difference in thermal conductivity of the three materials have significant effects on V only.

This implies that the dimensional accuracy of the scaffolds is more dependent on the SLM process parameters then the material properties. It is also observed that the strut dimensions along the build direction (V) have larger variation as compared to the strut dimensions parallel to the scanning direction (H_T). This means that the SLM process for lattice structures are difficult to control along the build direction, which may be due to a lack of support structures used in fabrication of the lattice structures. The insignificant difference between the strut dimensions for the TiTa, cpTi and Ti6Al4V lattice structures also lead to similar porosity of the lattice structures, as tabulated in Table 6.2.

Table 6.2 Compressive properties of SLM produced TiTa, Ti6Al4V and cpTi samples (n = 5)

Material	Porosity (%)	Elastic constant (GPa)	Yield strength (MPa)
TiTa	59.79 ± 0.68	4.57 ± 0.09	151.93 ± 8.47
Ti6Al4V	63.20 ± 0.55	5.47 ± 0.73	181.14 ± 15.05
cpTi	59.86 ± 0.59	4.29 ± 0.15	121.20 ± 3.67

6.3 Mechanical Characterisation

The fabricated cubic samples have designed dimensions of 10 mm by 10 mm by 10 mm, which are used as test coupons for compression tests based on ISO 13314:2011 (Mechanical testing of metals—Ductility testing—Compression test for porous and cellular metals).

Uni-axial compression tests were carried out, at room temperature (25 °C), to assess the compressive properties of the lattice structures, each with three replicates (n = 3), by using Instron Static Tester Series 5569 equipped with a 50 kN load cell. The loading speed was set at a constant of 0.6 mm/min, so as to maintain a constant strain rate for all tests as recommended by the standard. The compressive deformation rate has to be set such that the strain rate experienced by the samples are constant throughout.

The compressive deformation rate and strain rate is related by the following equation:

$$strain\,rate = \frac{compressive\,deformation\,rate}{height\,of\,test\,sample} \tag{6.2}$$

Using a standardized strain rate minimizes the differences in results obtained due to the effect of strain rate on hardening behaviour which may be present in some materials [5–8]. ISO 13314:2011 recommends a range of strain rates between 10^{-3} and 10^{-2}/s (0.06 and 0.6/min). To ensure consistency, the test samples are all strained at a uniform rate, without sudden changes, to minimize the effects of different strain rates [6, 7, 9]. The low strain rate of 10^{-3}/s is used as a high strain rate recommended by ISO 13314:2011 can lead to abnormal strain hardening behaviour in titanium [5–7].

The compression tests were carried out until axial deformation of the samples was equal to 100% or when the maximum loading of 50 kN was reached, whichever came first. The stress-strain curves, yield strengths and elastic constants in compression of the as-fabricated samples were then obtained through the compression tests.

The resulting elastic constant in compression and yield strength of the as-fabricated lattice structures are shown in Table 6.2. The slope of the straight-line portion of the stress-strain curve is established to define the elastic constant and the yield strength is taken as the stress at plastic compressive strain of 0.2%. The standard deviation in the elastic constant and yield strength may be due to the varying amount of powder adhesion on the struts which in turn, affects the compressive properties of the lattice structures.

The elastic constant of SLM produced TiTa lattice structures is lower compared to Ti6Al4V and is comparable to cpTi. The slightly higher TiTa elastic constant can be attributed to the presence of unmelted tantalum in the materials, resulting in resistance to the dislocation of the grains during compression. This results in stiffer material. The effect due to the presence of unmelted tantalum is more dominant in compression compared to tension where a lower modulus is obtained from TiTa compared to cpTi, as reported in Chap. 3. Nonetheless, TiTa still has the advantage of higher modulus to strength ratio as compared to cpTi in compression. In addition, TiTa also exhibit lower Young's modulus and higher strength compared to cpTi under tension. These make TiTa a more suitable material for use as porous and load bearing structures for biomedical applications where implants undergo both compression and tension. The results also coincide with the conclusions drawn in previous chapters.

6.4 Biocompatibility of Titanium-Tantalum

For the biocompatibility test, the overall dimensions of the lattice structures used are 10 mm in diameter and 2 mm in height. The generated CAD model and the as-fabricated SLM sample are shown in Fig. 6.6.

The human osteosarcoma cell line SAOS-2 was obtained American Type Culture Collection (HTB85, American Type Culture Collection, ATCC). The cells were cultured in McCoy's 5A modified medium (16600-082, Gibco, Life Technologies), supplemented with 15% fetal bovine serum and 1% antibiotics. The cells were cultured at 37 °C with 5% CO_2, routinely trypsinized after confluency, counted, and seeded onto the scaffolds. The scaffolds were sterilized by soaking in 70% ethanol for an hour and undergone autoclaving for 20 min at 121 °C. A cell suspension of

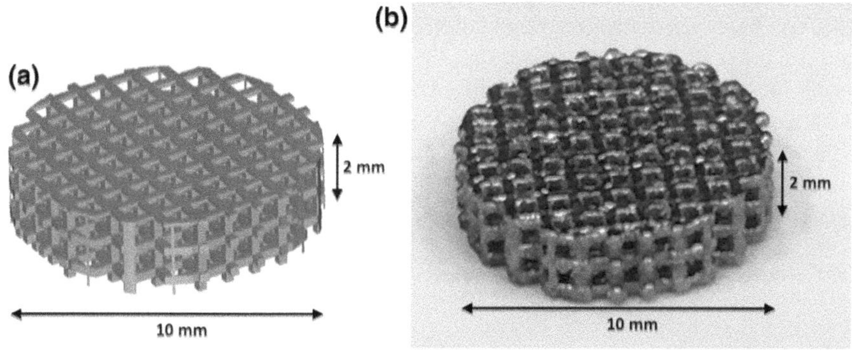

Fig. 6.6 Scaffolds for cell biocompatibility test **a** CAD design **b** SLM fabricated sample

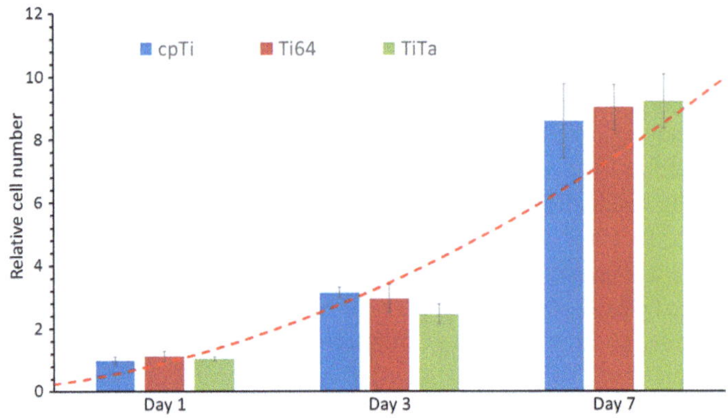

Fig. 6.7 Relative cell number on cpTi, Ti6Al4V and TiTa scaffolds

5×10^3 cells in 100 μL was added onto the top of each scaffold. After initial incubation of 3 h, 1 mL of culture medium was added to cover the scaffold.

To determine the biocompatibility of the scaffolds, tests using Quant-iT PicoGreen dsDNA Assay Kit (Life Technologies, Thermo Fisher Scientific) were performed on days 1, 3 and 7. Day 7 is the end of the culture period. The amount of double stranded deoxyribonucleic acid (dsDNA) corresponds to the number of SAOS-2 cells found within the scaffolds.

One-way analysis of variance (ANOVA) together with Tukey-Kramer post-hoc analysis was used to identify possible significant differences (with significant threshold: $p < 0.05$) between evaluate the TiTa, cpTi and Ti6Al4V.

Cell viability is assessed using dsDNA picogreen assay. The results are shown in Fig. 6.7.

dsDNA measurements show an increasing trend for all three materials ($n = 5$) as shown by the red dotted line in Fig. 6.7. After 7 days of in vitro culture, the SAOS-2 cells are found to be viable and proliferating on all the scaffolds. On day 1, 3 and 7, the relative cell number on the scaffolds of the three materials are similar. Using ANOVA, it is statistically determined that there is no significant difference between the results for all three materials on all the 3 time points. The slight difference in results can be attributed to the slight difference in porosity of the scaffolds due to the SLM process. The higher porosity allows more cell growth into the samples along the struts and better nutrient delivery and waste removal for cells that have migrated into the scaffolds [10, 11]. This factor becomes more critical as the number of days of in vitro testing increases.

6.5 Summary

This chapter shows that porous TiTa scaffolds fabricated using SLM are biocompatible. When benchmarked with cpTi and Ti6Al4V, the TiTa scaffolds show comparable biological results and manufacturability using SLM.

Based on the results, the key findings can be summarised:

(1) TiTa shows greater manufacturability using SLM when compared to pre-alloyed Ti6Al4V. This is shown by the higher dimensional accuracy obtained from TiTa lattice structures. It also produced scaffolds that have comparable dimensional accuracy compared to cpTi and the designed CAD.
(2) TiTa scaffolds showed high strength and lower elastic constant as compared to cpTi and Ti6Al4V.
(3) TiTa scaffolds showed similar biological response as cpTi and Ti6Al4V, making it a potential novel biocompatible material for biomedical applications.

References

1. A. Barbas, A.-S. Bonnet, P. Lipinski, R. Pesci, G. Dubois, Development and mechanical characterization of porous titanium bone substitutes. J. Mech. Behav. Biomed. Mater. **9**, 34–44 (2012)
2. A. Fukuda, M. Takemoto, T. Saito, S. Fujibayashi, M. Neo, D.K. Pattanayak, T. Matsushita, K. Sasaki, N. Nishida, T. Kokubo, T. Nakamura, Osteoinduction of porous Ti implants with a channel structure fabricated by selective laser melting. Acta Biomaterialia **7**, 2327–2336 (2011)
3. L. Mullen, R.C. Stamp, W.K. Brooks, E. Jones, C.J. Sutcliffe, Selective Laser Melting: a regular unit cell approach for the manufacture of porous, titanium, bone in-growth constructs, suitable for orthopedic applications. J. Biomed. Mater. Res. B: Appl. Biomater. **89**, 325–334 (2009)
4. L. Mullen, R.C. Stamp, P. Fox, E. Jones, C. Ngo, C.J. Sutcliffe, Selective laser melting: a unit cell approach for the manufacture of porous, titanium, bone in-growth constructs, suitable for orthopedic applications. II. Randomized structures. J. Biomed. Mater. Res. B: Appl. Biomater. **92**, 178–188 (2010)
5. Y.M. Wang, J.Y. Huang, T. Jiao, Y.T. Zhu, A.V. Hamza, Abnormal strain hardening in nanostructured titanium at high strain rates and large strains. J. Mater. Sci. **42**, 1751–1756 (2007)
6. D.R. Chichili, K.T. Ramesh, K.J. Hemker, The high strain-rate response of alpha-titanium: experiments, deformation mechanisms and modeling. Acta Materialia **46**, 1025–1043 (1998)
7. N.P. Gurao, R. Kapoor, S. Suwas, Deformation behavior of commercially pure titanium at extreme strain rates. Acta Materialia **59**, 3431–3446 (2011)
8. J. Zhang, H. Di, Y. Deng, R.D.K. Misra, Effect of martensite morphology and volume fraction on strain hardening and fracture behavior of martensite-ferrite dual phase steel. Mater. Sci. Eng. A **627**, 230–240 (2015)
9. S. Nemat-Nasser, W.G. Guo, J.Y. Cheng, Mechanical properties and deformation mechanisms of a commercially pure titanium. Acta Materialia **47**, 3705–3720 (1999)
10. C.M. Haslauer, J.C. Springer, O.L.A. Harrysson, E.G. Loboa, N.A. Monterio-Riviere, D.J. Marcellin-Little, In vitro biocompatilibity of titanium alloys dics made using direct metal fabrication. Med. Eng. Phys. **32**, 645–652 (2010)
11. S. Van Bael, Y.C. Chai, S. Truscello, M. Moesen, G. Kerckhofs, H. Van Oosterwyck, J.P. Kruth, J. Schrooten, The effect of pore geometry on the in vitro biological behavior of human periosteum-derived cells seeded on selective laser-melted Ti6Al4V bone scaffolds. Acta Biomaterialia **8**, 2824–2834 (2012)

Chapter 7
Future Trends in Titanium-Tantalum Formed by Selective Laser Melting

This chapter summarises all the work carried out for the completion of this thesis. Feasible future work regarding in-situ alloying of titanium-tantalum (TiTa) using selective laser melting (SLM) is also proposed.

7.1 In-Situ Alloying of Titanium-Tantalum using Selective Laser Melting

The formation of TiTa alloy using SLM by in-situ alloying for the first time is explored. Secondly, the microstructural and mechanical investigations are presented. Thirdly, the feasibility of TiTa scaffold fabrication using SLM is studied. Lastly, the biocompatibility of SLM fabricated scaffolds in vitro is summarised.

The key contributions of the research are elaborated in the following sections.

7.1.1 Formation of Bulk Titanium-Tantalum Using Selective Laser Melting

In order to develop a novel biomaterial, formation of TiTa by SLM is shown for the first time in this study.

Optimised Processing Parameters for Titanium-Tantalum

Work conducted on SLM is predominantly on pre-alloyed powder. In this study, the SLM capability in processing mixed powder is shown. The results have shown that SLM is capable in producing TiTa alloy with high density in a single step that was previously difficult to achieve. A processing window of SLM parameters is obtained for this in-house developed alloy with a set of optimised parameters identified using macrostructural analysis.

S. L. Sing, *Selective Laser Melting of Novel Titanium-Tantalum Alloy as Orthopaedic Biomaterial*, Springer Theses, https://doi.org/10.1007/978-981-13-2724-7_7

Characterised Microstructure of Titanium-Tantalum Formed By Selective Laser Melting

Microstructural analysis showed that the resulting TiTa alloy consist of β phase due to the stabilizing effect of β phase as a result of having tantalum as one of the alloying components and rapid solidification during SLM process.

Improved Mechanical Properties of Titanium-Tantalum Formed by Selective Laser Melting

The tensile test on bulk samples conducted concluded that TiTa showed a combination of high strength and lower Young's modulus compared to cpTi and Ti6Al4V parts produced using SLM. The SLM TiTa has Young's modulus of 75.77 ± 4.04 GPa, ultimate tensile strength of 924.64 ± 9.06 MPa and yield strength of 882.77 ± 19.60 MPa, shown in Table 4.2.

7.1.2 Formation of Lattice Structures Using Selective Laser Melting

In order to study the feasibility of using SLM to process TiTa for biomedical application, the new alloy is used to fabricate scaffolds by SLM.

Novel Statistical Modelling of Selective Laser melting for Lattice Structure Fabrication

Through statistical modelling, it is found that key SLM processing parameters, such as laser power, scanning speed and layer thickness, have significant effect on the dimensional accuracy and mechanical properties of the produced lattice structures. The strut dimensions of SLM fabricated lattice structures are found to be more sensitive to both laser power and layer thickness, as compared to laser scanning speed. Due to their effects on the strut dimensions, the mechanical properties of SLM fabricated lattice structures are all dependent on laser power, scanning speed and layer thickness.

Improved Manufacturability of Titanium-Tantalum Lattice Structure

TiTa shows greater manufacturability for lattice structure using SLM when compared to pre-alloyed Ti6Al4V. It also produced scaffolds that have comparable dimensional accuracy compared to cpTi. Likewise for the scaffolds, TiTa showed lower elastic constant as compared to Ti6Al4V scaffolds of the same design. However, the elastic constant of TiTa scaffolds are slightly higher than cpTi, due to the presence of unmelted tantalum. Nonetheless, TiTa scaffolds still have a higher strength to elastic constant ratio compared to cpTi. In addition, TiTa scaffolds also showed similar in vitro biological response as cpTi and Ti6Al4V.

The favourable results summarised above make TiTa a potential novel biocompatible material for future applications.

7.2 Future Work

The following future research directions are recommended:

7.2.1 *Development of Titanium-Tantalum Using Selective Laser Melting*

In this study, an alloying ratio of 50 wt% of tantalum to titanium is studied. For future work, lower alloying ratio of tantalum can be studied using SLM which may result in more favourable mechanical and biological response. Post-processing heat treatments can also be studied for TiTa.

7.2.2 *Thermal Modelling of Selective Laser Melting*

Current statistical modelling of fabrication of lattice structures using SLM has identified the key processing parameters in affecting the properties of the lattice structures. A possible extension to the study can include thermal modelling of thin struts or the whole three-dimensional (3D) lattice structures to confirm the effects. The thermal modelling can then be verified using the experimental results obtained in this study.

7.2.3 *In Vitro and In Vivo Biocompatibility Tests for Scaffolds Fabricated by Selective Laser Melting*

In vitro experiments in a static environment have been carried out in this project. Further biological response can be obtained using apatite forming ability test, live/dead viability cytotoxicity test and measuring the alkaline phosphatase activity. Dynamic environment can also be used as the behaviour of seeded cells on scaffolds is expected to be different in static and dynamic environment.

Histological studies can be carried to investigate the path taken by the cells in vitro to go into the depth of the scaffolds. In vivo tests in animal models can then be carried out with histological type works which will be useful in studying the cells and tissues attachment to the scaffold.

MIX
Papier aus verantwortungsvollen Quellen
Paper from responsible sources
FSC® C105338

If you have any concerns about our products,
you can contact us on
ProductSafety@springernature.com

In case Publisher is established outside the EU,
the EU authorized representative is:
Springer Nature Customer Service Center GmbH
Europaplatz 3, 69115 Heidelberg, Germany

Printed by Libri Plureos GmbH
in Hamburg, Germany